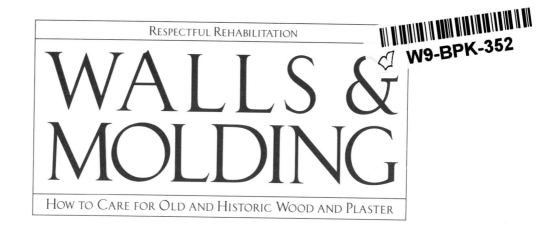

RESPECTFUL REHABILITATION

WALLS & MOLDING

HOW TO CARE FOR OLD AND HISTORIC WOOD AND PLASTER

WALLS & MOLDING

HOW TO CARE FOR OLD AND HISTORIC WOOD AND PLASTER

NATALIE SHIVERS

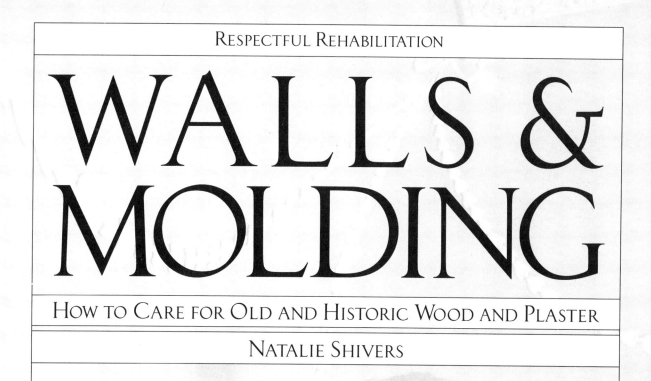

PRESERVATION PRESS   JOHN WILEY & SONS, INC.

Library of Congress Cataloging in Publication Data

Shivers, Natalie W. (Natalie Wilkins)
    Walls & molding : how to care for old and historic wood
    and plaster / Natalie Shivers.
        p.   cm.   — (Respectful rehabilitation)
    Includes bibliographical references.
    ISBN 0-471-14432-0
        1. Interior walls — Maintenance and repair.   2. Moldings —
    Maintenance and repair.   3. Historic buildings — Maintenance
    and repair.   I. Title.   II. Title: Walls and molding.   III. Series.
    TH2239.S45   1990
    693'.6 — dc20   89-78511

Printed in the United States of America
                            5 4

Designed by Meadows & Wiser, Washington, D.C.
Unless otherwise indicated, drawings by Mathew Barac, IMG
Publishing, New York City.

Cover: Conservator Matthew Mosca using a toothbrush to remove
paint from plasterwork in the large dining room at Mount Vernon.
(Courtesy of The Mount Vernon Ladies' Association)

CONTENTS

Unpainted wood ornament
at Drayton Hall, 1738–42,
Charleston, S.C. (NTHP)

PREFACE

Plaster walls and molding, woodwork, stenciling and painted finishes—technically, these are "nonfunctional" components of old buildings. They are not required to hold up structures, keep out the rain or heat or ventilate rooms. Yet these architectural elements and finishes serve a vital role in giving old buildings much of their visual character. The tactile qualities of old plaster, the colors of painted finishes, the patterns of ornamental wood and plaster molding all give texture and proportion to a room. Their presence is easy to take for granted, yet they are sorely missed when absent. Unfortunately, it is sometimes not until a plaster wall is replaced with one of gypsum board, or a plaster ceiling medallion is removed and the surface left bare, or "mahogany" graining is stripped from a wood door that we realize just how important these elements are.

Plaster, wood and paint are subject to deterioration generated from sources on both sides of walls and ceilings. Not only are they constantly assaulted in the course of people's daily activities, but they also quickly display evidence of unseen moisture, structural and mechanical problems. Almost by definition, old buildings have loose or cracked plaster, damaged or missing moldings, and degraded finishes. In some cases, these problems are signs of more serious troubles underneath and need to be investigated further before proceeding with rehabilitation of the surface. More often, however, plaster, woodwork and paint can be patched and repaired or revived without wholesale removal and replacement.

This book is intended for people interested in preserving interior finishes and architectural ornament in old and historic buildings. The advice and information here

Wood molding from Drayton Hall, 1738–42, Charleston, S.C. The patterns are modillions with acanthus, egg-and-dart and bead-and-reel. (NTHP)

Opposite: Library, Hutzler House, Baltimore, a late 19th-century residence. (Ella Hutzler Oppenheim)

7

Double staircase at Waverly Plantation, c. 1852, Columbus, Miss. (Jack Boucher, HABS)

pertain to all old buildings, especially those more than 40 years old, whether of special historical or architectural value or not. The text presumes that any necessary structural and mechanical repairs have been made and you are now ready to deal with the visible surfaces and their ornamentation.

Some of this work can be done by a layperson who has the interest and necessary skills, using the basic methods and materials described in the text and illustrations. Commercial methods for complex and large-scale projects are also discussed. If you hire a contractor, the text will help you define the scope of work, evaluate the options for correcting specific problems and monitor the job.

The methods and materials described here are aimed at rehabilitating rather than restoring historic structures to an earlier time and appearance. Much of the information is based on the principles expressed in the Secretary of the Interior's Standards for Rehabilitation and Guidelines for Rehabilitating Historic Buildings. Recommendations given here, however, do not represent specific Interior Department policies, and some techniques discussed may not be acceptable to the department's National Park Service regional offices for receipt of the federal rehabilitation tax credit, awarded to income-producing historic properties. Consult directly with the National Park Service or your state historic preservation office if you intend to apply for the tax credit.

NATALIE SHIVERS

ACKNOWLEDGMENTS

This book is a result of my own experience rehabilitating the Los Angeles Central Library and the Bradbury Building, both case studies in repair and rehabilitation problems. In addition, publications such as *The Old-House Journal*, *Association for Preservation Technology Bulletin*, *Fine Woodworking*, *Historic Preservation* and *Preservation Briefs* published by the National Park Service, notably that by Marylee MacDonald, proved indispensable in providing further valuable information about historic wood, plaster and painted finishes with solutions. Specific references to these useful sources are included in the bibliography in the back along with other resources.

Cornice showing Federal influence with dentils and the suggestion of a garland in the frieze. (HABS)

Professional conservators, restoration architects and contractors helpfully reviewed parts of the text and consulted on particular issues. They included William Adair, Gold Leaf Studios, Washington, D.C.; Ken Breisch, Southern California Institute of Architecture, Santa Monica; Brian Considine, The Getty Museum, Santa Monica; Bill Gilliland, architect, Los Angeles; Tony Heinsbergen, A. T. Heinsbergen and Company, Los Angeles; H. Ward Jandl and Kay Weeks, Technical Preservation Services, National Park Service; John Leeke, preservation consultant, Sanford, Maine; Marylee MacDonald, Small Homes Council, Champaign, Ill.; William MacMillen, Richmondtown Restoration, Staten Island, N.Y.; Roger H. Moss, The Athenaeum, Philadelphia; Robert Mussey, Society for the Preservation of New England Antiquities; Ray Pepi and Jay Cardinal, Building Conservation Associates, New York; Edward Pinson and Debra Ware, Pinson and Ware, Los Angeles; John Twilly, Los Angeles County Museum of Art; Martin Weil, restoration architect, Los Angeles; Frank Welsh, paint conservator, Bryn Mawr, Pa.; Rosamond Westmoreland, paint conservator, Los Angeles; and Gail Caskey Winkler, LCA Associates, Philadelphia.

I want especially to thank Diane Maddex and Janet Walker of The Preservation Press of the National Trust for Historic Preservation and Angela Miller, Sharon Squibb and IMG Publishing whose support made the book a reality. Also, I am indebted to my employers Norman Pfeiffer and Brenda Levin both for the chance to work on exceptional historic buildings and for their forbearance, patience and tolerance while I wrote this book. Lastly, I am grateful to my in-house editorial adviser, my father, Frank R. Shivers, Jr., for help with illustrations and editing.

Opposite: Master bedroom, Frank Lloyd Wright Home and Studio, Oak Park, Ill., restored to the year 1909, Wright's last year there. Ten coats of paint were removed from the mural, which has been meticulously restored.(Jon Miller, Hedrich-Blessing, for the Frank Lloyd Wright Home and Studio Foundation)

AMERICAN INTERIOR STYLES

When European settlers established the American colonies, they brought with them traditional architectural forms as varied as their origins. Sources for early vernacular architecture in the new country included a mixture of medieval and Renaissance building traditions and styles imported from the immigrants' native countries to the geographic regions where they settled: the British on the eastern seaboard; Spanish in areas that are now Florida, New Mexico, Arizona, Texas and California; French in the St. Lawrence and Mississippi River valleys; Dutch in the Hudson River and Mohawk valleys; and German in Pennsylvania. These traditions were adapted to each region's unique conditions, including climate, local building materials, and political and economic circumstances.

As the British established political dominance in the eastern colonies, their cultural influence prevailed. Their introduction of Renaissance, or classical, styles in the early 18th century had a profound impact on colonial architecture that spread across the country as settlers moved west. Other traditions continued to flourish, particularly in areas remote from the British colonies. However, from the 18th through the mid-19th centuries, the classical buildings of ancient Greece and Rome inspired the forms and details of a large number of structures, particularly public buildings and houses for the wealthy. America's interest in classically based designs continued with Colonial Revival styles, which were popular through the middle of the 20th century.

Opposite: Victorian interior with richly patterned wall surfaces, stair railing and furnishings. (Library of Congress)

Detail of a tabby wall. Made of lime, sand, water and shells, such walls were the oldest in the Spanish-settled Southeast. (Walker Evans, Library of Congress)

Because of the continuing impact of classical principles on American architecture, any discussion of wall treatments in American interiors must begin with the classical orders and building styles. Although many styles and traditions were not derived from classical prototypes, the orders provided a system of organization and ornament that was consistently returned to in one form or another throughout the history of American interiors.

There were five classical orders, three established in ancient Greece — Doric, Ionic and Corinthian — and two

Capitals and entablatures of the classical orders as depicted in Andrea Palladio's *The Four Books of Architecture*.

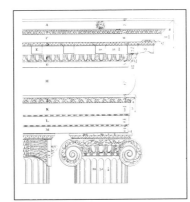

Ionic (Greek)

Tuscan (Roman)

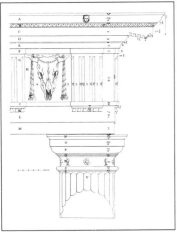

Doric (Greek)

Corinthian (Greek)

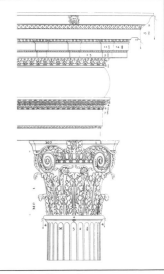

Composite (Roman)

added by the Romans — Tuscan and Composite. The proportions and details of each order were governed by specific sets of rules. Proportions of each element of a column, for example, were based on the diameter of the column base as a unit of measure. Each order also generated a system of ornament that controlled both the character and disposition of moldings.

European neoclassical architects and theorists in the 15th through 18th centuries later established their own rules governing the proportions, details and use of classical elements. They identified and reinterpreted classical models based on a series of Renaissance treatises such as Andrea Palladio's *I Quattro Libri dell'Architettura* (1570, Venice; 1715, England). Principles developed by these theoreticians gradually entered into the consciousness of American artisans through high-style examples, builders' guides and the influence of their peers immigrating from Europe.

Until the middle of the 19th century, few professional architects worked in America. Before that and even through the late 1800s, artisans and builders usually were responsible for the design of buildings. Until they began to lose their autonomy, most builders were familiar with — and able to design and execute — the elements of classical architecture. Builders' guides such as Batty Langley's *City and Country Builder's and Workman's Treasury of Designs* (England, 1740) and Asher Benjamin's *The American Builder's Companion* (Boston, 1806) described how to draw and execute moldings. While artisans were expected to know the basic rules of proper proportion, ornamentation and assembly of individual building components, and to extrapolate from those rules as circumstances demanded, they frequently took liberties with their models. Exact reproduction of classical ornament was more the exception than the rule in 18th-century America, rarely seen on any but the most important public and residential buildings.

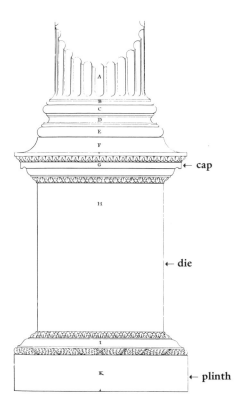

Classical pedestal from Andrea Palladio's *The Four Books of Architecture.* Its subdivisions correspond to the parts of a lower interior wall (chair rail, dado, base).

WALLS

In American interiors of many styles, walls were divided into three parts:

- dado, the lower wall above the baseboard
- field, the middle wall between the dado and cornice

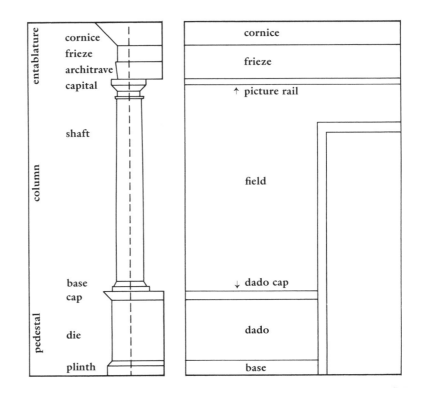

Corresponding parts of an interior wall and a classical order.

"Since mouldings do, as it were, compose the alphabet of architecture, and that without a perfect knowledge of their several attributions and combinations, it is impossible to acquire any proficiency, their uses and shapes should be well considered...."

— Stephen Riou, *The Grecian Orders of Architecture*, 1728

- cornice, the uppermost part of the wall at the juncture of wall and ceiling.

This practice was derived from the classical order: the parts of a wall corresponded to the parts of a column. The dado is the equivalent of a column's pedestal, the field its shaft, and the cornice its capital. Within these major divisions, walls were further subdivided to correspond to column subdivisions. For instance, a wall's baseboard corresponds to the plinth, or squared base, of a column, and the dado cap represents the column's surbase.

MOLDING

Molding is a nonstructural building part that ornaments primary structural elements and creates varied effects of light and shadow. Molding in classical architecture was the subservient part of an order. It appeared at the junctures of the plinth, shaft, capital, entablature and pediment. Ancient Greeks used molding in their buildings to divide surfaces into smaller parts, creating visual interest with highlights and shadows. Romans simplified and reduced the quantity of Greek moldings, using mechanical tools and compasses to reproduce it. Most Roman moldings had circular profiles, while Greek moldings generally had profiles based on the ellipse, parabola or hyperbola.

The eight classical shapes of molding were the fillet, astragal (bead), torus, scotia, ovolo, cavetto (cove), cymatium (cyma recta or ogee) and inverted cymatium (cyma reversa or reverse ogee). Fillets, astragals, torii, ovoli, scotias and cavetti are considered simple moldings because they are formed from a continuous curve such as the arc of a circle. Complex moldings (the ogee and reverse ogee) have irregular curvatures and were often combined with simple moldings.

Decorative motifs for molding were specific to each profile and its visual function. The egg-and-dart pattern, for instance, was used for ovoli, and astragals generally were ornamented with a bead-and-reel pattern. Other classical patterns such as guilloche, paterae and acanthus leaves also had specific applications.

Molding in American interiors

Depending on the scale and style of a building, molding could adorn most interior architectural elements: wall and door panels; door and window frames; window mullions; the tops, bottoms and corners of walls; dadoes; fireplaces and chimney breasts; handrails and stairs. In simple buildings cost would have limited the general use of elaborate molding until it began to be produced by machines in the third quarter of the 19th century. Ornamentation cost money, and the extent of its use and stylishness usually reflected the wealth and taste of a building's owner. In a sense, after 1850, decoration became more democratic: the manufacture of ornament by machines allowed it to become affordable for an increasing number of people. However, some vernacular buildings, such as those of Spanish origin in the Southwest, rarely had decorative molding of any kind.

Where molding was used, it often had practical value as well as ornamental appeal. The baseboard, for example, covered the joint between the floor and the wall and protected the wall as the floor was being swept or mopped. In fine buildings the dado, sometimes paneled, was used to place

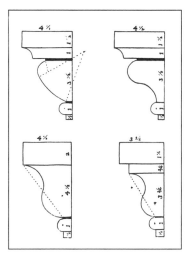

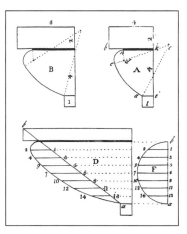

Roman (top) and Greek (bottom) profiles from Asher Benjamin's *The American Builder's Companion,* an early 19th-century builder's guide.

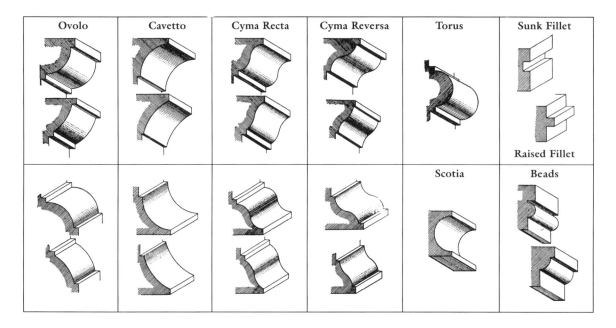

Ovolo	Cavetto	Cyma Recta	Cyma Reversa	Torus	Sunk Fillet
					Raised Fillet
				Scotia	Beads

Basic classical molding profiles from William Ware's *The American Vignola*. Such profiles have been the "building blocks" of interior ornament for centuries.

decoration above the height of furniture, at eye level. It also protected the wall against damage by chairbacks and provided a permanent and ornamental back to seats and benches without backs. The dado cap (later known as the chair rail) covered the joint where the different materials of dado and field met, and the cornice covered the joint between the ceiling and the wall. The field was the area designated for decoration, as it was fully visible and protected from chair scrapes. Molding separating the frieze at the top of the wall from the field usually served as a picture rail after the middle of the 19th century.

Dating molding

Moldings generally had the same or similar profiles whether made by hand or machine, of plaster, wood, cast plaster or various plaster or wood-substitute materials such as tin. Profile variations evolved, however, to suit specific architectural and decorating styles and as developing technology allowed richer and more complex ornamentation.

Restoration architects and architectural historians familiar with the development of local technology and styles sometimes can date molding. Evidence for dating includes the molding profile, type of nails used to secure it (hand wrought or machine cut) and method of manufacture

(hand plane, circular saw or band saw, for instance). The process of dating molding, however, requires a well-trained eye, because there are many variables and few hard and fast rules that apply consistently throughout the country.

How wood molding was made

Until the mid-19th century, molding was made with hand planes, chisels and gouges. Carpenters would carry their tools and make most wood moldings on site. Wide pieces of trim such as crown moldings were made in woodworking shops, where large planes were pulled by several apprentices and guided by a master artisan.

The earliest known planing machine in America was patented in 1828 but was not adapted for molding manufacture until the 1840s. By the middle of the 19th century, molding machines of various kinds were able to produce ornament on a whole new scale. This new machinery could do the work of dozens of joiners to produce large quantities of molding of the exact same profile at a low cost. With the establishment of planing mills using water- and steam-powered machines, the manufacture of decorative elements was soon widespread, making elaborate and sophisticated ornament accessible to all.

While the design and construction of moldings were a necessity as well as a point of pride for carpenters of the

Left: Making molding with a hand plane. (John Leeke)

Right: 18th-century hand plane. (Michael Devonshire)

Early 19th-century interior showing classical influence in the wall and fireplace design. (NTHP)

Machine-made moldings from A. J. Bicknell's *Detail, Cottage and Constructive Architecture*. Wood moldings generally were made in sections assembled to make complex profiles.

18th and early 19th centuries, by 1860 builders and carpenters had become assemblers of profiles from stock molding. Pattern books such as Asher Benjamin's *The American Builder's Companion* (1806 and following) gave way to manuals such as A. J. Bicknell's *Detail, Cottage and Constructive Architecture* (1873) describing how to assemble prefabricated moldings.

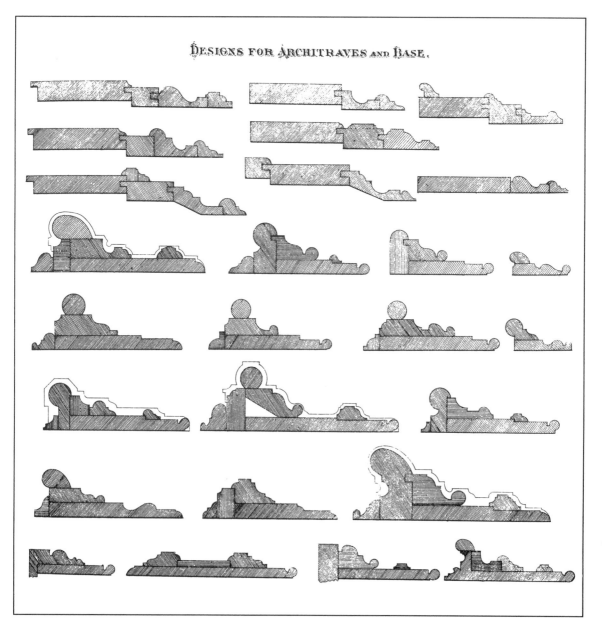

DESIGNS FOR ARCHITRAVES AND BASE.

The 20 years after the Civil War saw the invention and patenting of many machines and methods to produce decorative wood and metal ornament. Newly developed saws, planes, boring and sticking machines, joiners, shapers, mortise-and-tenon machines, lathes, sanders and presses mass-produced an increasing range and variety of ornament. As machines became capable of heavier work, ornament increased in complexity and relief. Thus, moldings in Classical Revival buildings of the early 19th century generally consisted of simply trimmed boards while moldings of the second half of the 19th century usually had more complex, heavier, three-dimensional profiles.

Around 1880 moldings began to be identified using a standard numbering system, such as the "4000 series" or "5000 series." The current WP series of standard moldings was instituted in 1957 by the Western Wood Moulding and Millwork Producers. WP numbers describe both the shape and size of profiles, such as WP-47, which is $1\frac{1}{16}$- by $4\frac{5}{8}$-inch crown molding.

"The guide and masterpiece of all architecture depends solely on the magnitude and composition of mouldings, these are the touches of art which give force and beauty to whatever is intended...."
— Thomas Skaife, *A Key to Civil Architecture*, 1776.

WOODWORK

Interior woodwork is variously known as trim, millwork and joinery and can include these elements: baseboards, chair rails, picture moldings, cornices and entablatures, and door and window trim, as well as doors and windows, dadoes, wainscots, paneling, mantels and overmantels, and built-in cabinets and niches.

COLONIAL (1600–1720)

Walls in the 17th and early 18th centuries were typically finished with vertical or horizontal boarding or plaster (see Plaster in this chapter), applied to the interior framing system between the structural posts and beams. Structural members such as summer beams, joists and posts were left exposed and sometimes ornamented with chamfered corners or quirked beads.

Partial- or full-height wooden wainscots have been among the most popular wall treatments used in American

21

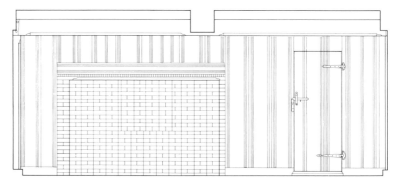

Typical 17th-century interiors. Salem, Mass., dining room (Library of Congress) and elevation from the Cushing House, Hingham, Mass., both with vertical board sheathing on the fireplace wall, exposed structural posts and beams and few or no ornamental moldings around openings.

interiors from the 17th century to the present. Brought to this country from England, where wainscots came into general use in the Middle Ages, they gave 17th-century American interiors a decidedly dark and heavy medieval character.

Boards were of equal or random width, generally overlapped, tongue-and-grooved or half-lapped. Early ways of ornamenting the sheathing included creased molding at the middle or edges of boards, simple beads and feather-edged joints.

The aesthetics of wooden wainscots may have been secondary to their function in saving energy, helping to keep walls weathertight and rooms warm. In fact, the 17th-century Massachusetts Bay colonists made a distinction between woodwork installed for warmth and woodwork used for ornamental purposes. The latter was considered an indulgence, difficult to justify in a colony where, as one citizen proclaimed in 1620, "Our purpose is to build for the present such houses as, if need be, we may with little grief set afire and run away by the light. Our riches shall not be in pomp but in strength."

In 17th- and 18th-century buildings, the fireplace wall frequently was sheathed with wooden boards and, in later buildings, with panels. These covered the juncture between the masonry fireplace and chimney breast and the wood frame wall in which they were set, while still allowing heat from the flue to radiate into the room. Early mantelpieces were rare. Often the horizontal lintel and masonry piers were left exposed. Occasionally, the lintel was covered with a horizontal board, sometimes molded or with incised decoration.

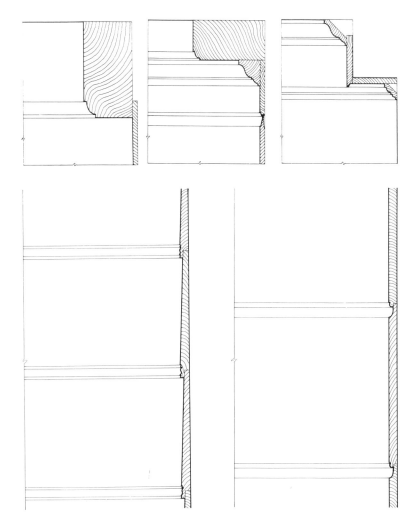

Top: Sections and partial elevations of colonial wood cornices. Typical cornices ranged from structural beams with molded edges (left) to applied moldings (right), sometimes combining both (center).

Center: Sections and partial elevations of colonial horizontal board wainscoting showing enrichment of half-lapped boards with molded edges.

Below: Plans and partial elevations of colonial vertical board wainscoting showing a half-lapped joint with molded edges (left), butted joint with batten or applied molding (center) and tongue-and-groove joint with raised panel (right).

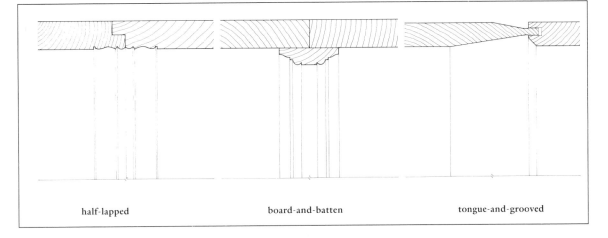

half-lapped   board-and-batten   tongue-and-grooved

Above: Georgian interior where early classical influence has led to pilasters, raised wall panels and a built-in cupboard.

Right: Plans and partial elevations of 18th-century raised paneling showing stiles and rails with molded edges (left) and applied moldings (right).

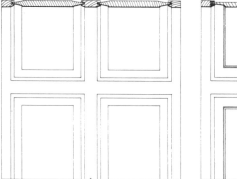
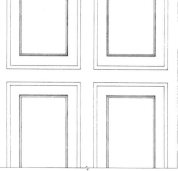

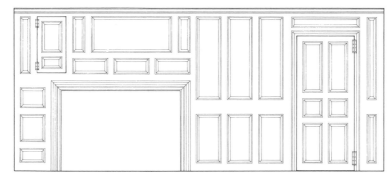

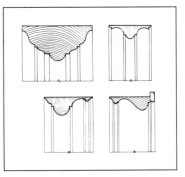

In some places, such as the Massachusetts Bay Colony in the early 18th century, horizontal boarding was commonly used on all walls, as it required less labor and thus less cost to trim boards and secure them to the frame than to cut and nail laths and prepare and apply plaster.

Typical ceiling treatments consisted of leaving the joists and floorboards exposed above, although in some finer buildings plaster ceilings were installed between the joists.

GEORGIAN (1700–80)

The aesthetics of interiors were determined more by construction than ornament until the beginning of the 18th century. At that time, however, decorative forms and moldings gained more attention. The casing or boxing-in of the house frame was one the earliest expressions of Renaissance influence. Applied moldings also became more common — for instance, around fireplace openings and at the junctures of walls and ceilings.

In formal Georgian interiors horizontal and vertical board wainscots lost favor to the more stylish wood paneling. Fine Georgian paneling generally consisted of raised panels with beveled edges, set within frames of stiles and rails with either quarter-round molding run along their edges or, sometimes, with applied bolection molding. Rails and stiles generally were mortised and tenoned together and secured with wood pegs.

In the course of the 18th century wood paneling gradually became classicized as the result of strong English influence on American builders. Organization of walls

Left: Elevation of early 18th-century interior with typical features, including raised wood paneling, wood casing and moldings hiding structural posts and beams and applied moldings around the door, window and fireplace openings.

Right: Bolection moldings, popular around fireplaces, doors and windows as well as panels in 18th-century interiors.

Elevation of a high-style Georgian interior, the southwest room at Carters Grove, James River, Va., c. 1740, showing classical influence in symmetry, tripartite division of the wall surface and literal use of classical elements such as pilasters, triglyphs, dentils and crosette trim around the fireplace.

into three parts and a classical sense of symmetry and balance signaled the new trend. Initially, though, there was not much direct copying of Greek or Roman detail, nor was much thought given to the "proper" scale and proportion of elements or even to the overall composition of a wall.

Paneled walls, wainscots and ornamental chimney breasts were features of many formal Georgian interiors. The fireplace wall was usually paneled, while the three other walls were either paneled or plastered. Generally, the fireplace wall was the most highly ornamented, with bolection moldings around the fireplace and overmantel panels. The cornice was generally simple and without explicitly classical details. Ceiling treatments usually consisted of boxing in the structural beams supporting the floor above.

Around the second quarter of the 18th century (the dates varied in different parts of the country), actual classi-

cal forms — columns, pilasters, entablatures, cornices with classical motifs — became popular ways of ornamenting formal interiors. Walls began to be conceived as large-scale compositions, with an overall order and balance. Special attention was given to the decoration of openings: doors and windows, fireplaces and built-in cupboards. In most formal interiors of this time, the same or similar moldings generally were used around these openings as well as on wall panels.

Formal rooms in more elaborate buildings sometimes were decorated with complete classical orders of pilasters and entablatures. There were obvious borrowings from classical sources, such as acanthus leaves, consoles and architraves of plain or broken pediments adorning window and door openings and the chimney overmantel. Even simpler buildings adopted classical designs in the form of architrave trim, pilasters, cornices and paneled walls. Small houses built at this time may have had only plaster walls with a minimum of ornamental woodwork, however.

Moldings of the 18th century generally followed pattern books inspired by classical antiquity. Most were based on Roman profiles and composed essentially of segments of a circle, such as ogees and beads. They tended to have heavy, bold profiles and often comprised simple or complex moldings used as individual elements.

FEDERAL (1780–1820)

The evolution of molding profiles reflected, to a great extent, contemporary stylistic trends. One major influence toward the end of the 18th century was the work of Scottish designers Robert and James Adam, whose designs became well known through such books as *The Works in Architecture of Robert and James Adam* (1773). They favored smooth plaster walls ornamented with delicate molded decoration of classical derivation. Fashionable American interiors of the Federal era followed suit.

Paneling for entire walls lost favor, although paneled dadoes and fireplace walls remained common. Paneling was still used, at least in New England houses, for finishing the walls of back rooms, particularly the kitchen and the less important rooms of the second floor.

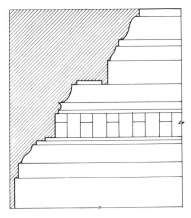
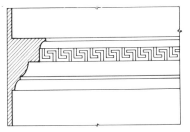
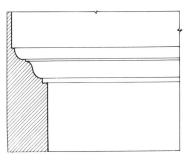

From Carters Grove, sections and partial elevations of a cornice (top), chair rail (center) and baseboard (bottom). Molding profiles and surface ornament such as the cornice dentils and chair rail fret are derived from classical forms.

27

Parlor, Carroll Mansion, Baltimore, c. 1808–18, a Federal interior. The marble mantel and chandelier are later additions. (M. E. Warren)

Plaster walls were ornamented with a baseboard, chair rail and cornice that continued around all four walls of a room. Baseboards were used to finish the wall against the floor. Early baseboards often were set so that their surfaces were flush with the plaster. Later, baseboards generally projected beyond the face of the plaster and may have had carved or applied molding. Finely detailed composition, or putty, ornament that looked like carved wood often adorned mantelpieces, door and window pediments, entablatures and cornices.

Throughout the Federal period, classical models for molding were further refined. Because artisans were working in wood rather than stone, as ancient Greeks and Romans had done, they exploited its capabilities and created moldings with thinner edges and flatter projections. At the same time, fashion favored lightness and delicacy of ornament. Molding forms changed from profiles adapted from the simpler orders to designs of a finer scale with delicate beading and elliptical shapes.

GREEK REVIVAL (1820–60)

The Greek Revival style represented a conscious return to Greek forms; by this time, classical forms were believed to have originated in Greece rather than Rome. In addition, the architecture of the Greek city-states where liberty

Elevation of the drawing room, Imlay House, Allentown, N.J., a transitional Federal interior. Typical features of high-style interiors of the early 19th century included plaster walls above paneled dadoes, classically influenced moldings and elaborate treatments of the fireplace opening and chimney breast with a classical pediment and pilasters.

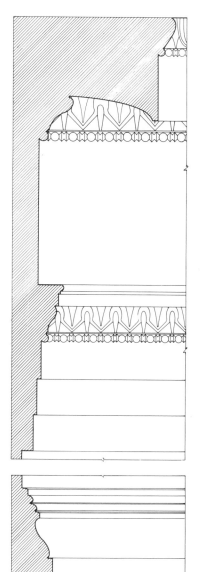

flourished, rather than that of the Roman Empire, with its negative imperial associations, seemed a more suitable ideological model for the United States. Significantly, the sixth edition of Asher Benjamin's *The American Builder's Companion* (1827) presented drawings of the Greek orders for the first time.

Parts of each order, including the entablature, were larger than their Federal counterparts, conveying a sense of solidity and simplicity. Profiles generally were composed of segments of the ellipse, which were thought to create more pleasing reflections of light than segments of a circle. In rural areas, flat boards were sometimes used instead of molded trim.

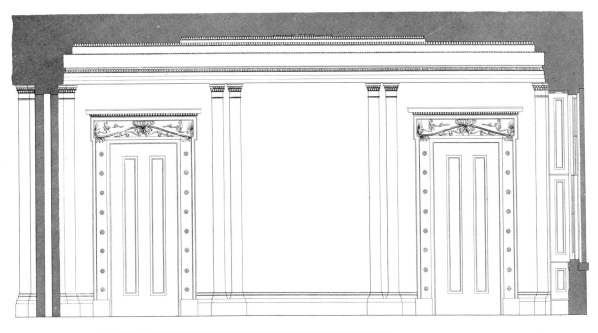

Victorian (1840–1900)

The Victorian era heralded a period of stylistic eclecticism. The list of styles that prominent architect Alexander J. Davis could supply his clients in the 1850s and 1860s included "American Log Cabin, Farm Villa, English Cottage, Collegiate Gothic, Manor House, French Suburban, Switz Chalet, Switz Mansion, Lombard Italian, Tuscan from Pliny's Villa at Ostia, Ancient Etruscan, Suburban Greek, Oriental, Moorish, Round Castellated." Styles varied widely and changed quickly, and interiors, particularly of more public rooms, were redecorated with some frequency.

The prevailing style was usually reflected in mantelpieces, molding profiles, decorative motifs and wall treatments — elements that were easily changed with new fashions, particularly with the availability of machine-made ornament. Gothic Revival profiles, for instance, often displayed pointed beads and deeply undercut circular and elliptical shapes. Other popular features of the Gothic style included dark paneling (cheaper woods were grained for a dark hardwood effect if necessary), painted imitation stone walls, particularly in hallways, and tapestry or embossed leatherlike wall coverings.

Above: Elevation of a Greek Revival interior from Minard Lafever's *A Modern Builder's Guide*, 1833, characterized by monumental proportions, enlarged and simplified ornament and the use of ancient Greek rather than Roman motifs.

Opposite, left: High-style Greek Revival cornice and base from Minard Lafever's 1833 guide.

Opposite, right: Parlor, Waterloo Row, Baltimore, designed by Robert Mills in 1816. This Greek Revival row house interior displays characteristic monumental proportions and classically influenced molding and ornament. (Baltimore Museum of Art)

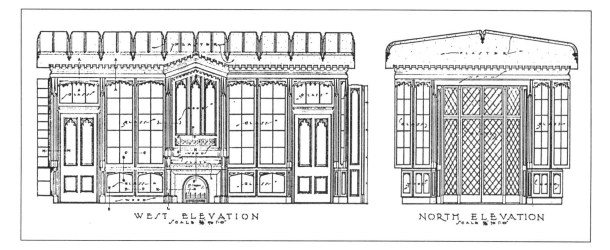

WEST ELEVATION

NORTH ELEVATION

Above: West and north elevations of the library at Riverside, Burlington, N.J., a Gothic Revival interior designed by John Notman in 1839. (Ernest L. Ergood, HABS)

Right: Victorian interior with tripartite wall division using paneled wainscot, wallpapered field and molded cornice. (Library of Congress)

Interiors of the late 19th century were strongly influenced by such reform-minded critics as Charles Eastlake, author of the popular *Hints on Household Taste in Furniture, Upholstery & Other Details* (first American edition, 1872). Eastlake condemned the novelty and what he

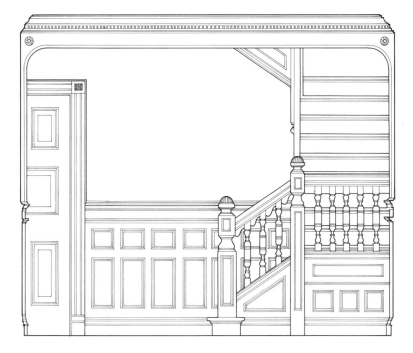

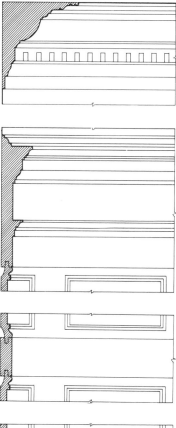

saw as dishonesty of the many revival styles popular at that time. His influence helped revive the popularity of the tripartite wall division (dado or wainscot, field, frieze or cornice).

Custom-designed wood paneling was expensive and generally appeared only in public areas and in entry halls and dining rooms of larger middle-income homes, where it would achieve the maximum exposure. But by the 1880s prefabricated wood paneling, glued onto heavy cloth for easy installation, was widely available. The cheapest version was of plain vertical boards (sometimes of two woods to look custom-made) with a wooden cap. The tripartite wall treatment was also achieved by the use of moldings and different colors of paint or wallpaper.

Picture rails set just below the frieze also were generally recommended to avoid damaging plaster walls with nail holes and to allow flexibility in hanging pictures. Modest houses often confined their wall ornament to a picture rail, placed against the ceiling to double as a cornice. More formal rooms had cornice moldings as well, sometimes with the picture rail as the bottom part.

At the end of the 19th century two divergent move-

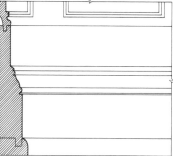

Top left: Elevation of an Eastlake-inspired interior.

Above, top to bottom: Section and partial elevations of a Victorian cornice, chair rail, wainscot and base.

Left: Dining room, Melville Klauber House, a Craftsman house in San Diego, designed by Irving Gill. (Marvin Rand, HABS)

Right: Elevation of a Craftsman interior.

Opposite: Colonial Revival interior, 1914. (Library of Congress)

ments emerged: one a continuation of the eclectic revival styles and the other a reform effort generating the Craftsman and Colonial Revival styles.

CRAFTSMAN (1895–1940)

Craftsman interiors were based on principles espoused by the widely influential monthly magazine *The Craftsman* (1901–16), published by Gustav Stickley, an architect, furniture and interior designer, and furniture factory owner. Stickley championed the Arts and Crafts aesthetic, popularized in mid-19th-century England by John Ruskin and William Morris. This movement rejected superfluous ornament and the "meaningless" excesses of the revival styles. It advocated instead the honest expression of structure and the straightforward use of materials.

As a result, Craftsman interiors derived their character from joinery. Structural elements such as exposed posts and ceiling beams were treated as ornament. Walls generally were divided into three parts: wainscot, field and frieze, the latter usually delineated by a picture rail but without a cornice. Wainscots, in particular, were thought to give a sense of "friendliness, mellowness, and permanence." Craftsman wainscoting generally consisted of single vertical panels, board-and-batten paneling or a skeleton of vertical wooden strips dividing the plaster wall into panels, with a cap rail that was sometimes used as a plate rail. Another treatment provided niches at the tops of panels for pottery. Wainscot heights of five feet or more were typical for these interiors.

Both the extent of exposed woodwork and the many

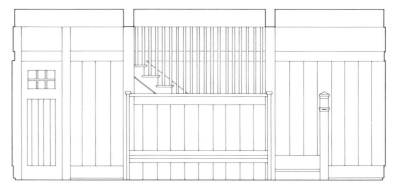

built-in pieces of furniture such as window seats and side-boards contributed to the distinctive handcrafted appearance, although machines were responsible for all but the final finishing and assembly of parts.

REVIVAL STYLES (1880–1955)

Colonial Revival styles included early French, Dutch and Spanish as well as early American colonial. American Colonial Revival began with the classical detailing of the Queen Anne style in the 1870s and gained widespread popularity in the 1890s. As with Craftsman interiors, simplicity and the return to essentials — or America's roots — were advocated. A range of models from New England saltboxes to elaborate Georgian buildings served as sources for the styles.

Walls were sometimes paneled floor-to-ceiling or divided into two parts, field and frieze. Georgian Revival interiors were ornamented with simple moldings and classical motifs such as dentils, garlands and swags and sometimes featured columns and pilasters as well. Manufacturers even produced elements based on regional colonial details, such as Tidewater wainscots.

In the 1920s full wall paneling, considered "colonial," became fashionable for interior rooms in three forms: (1) hardwood panels with stiles and rails, (2) pine paneling of knotty wood and (3) veneered plywood panels set off by hardwood strips. A cheaper method was to divide a plaster wall into panels merely by adding vertical strips of moldings. Even

"No machine ever invented or introduced to the public notice is capable of doing so much to benefit the million. Beautiful and graceful forms may now adorn their dwellings as well as the palaces of the rich and great . . . . It is useless to carry segments of circles to the turner's shop, or keep carvers at work upon gothic molding. This machine, in such work, has the capacity of thirty first-class mechanics."

— Catalogue description of the spindle shaper, 1868

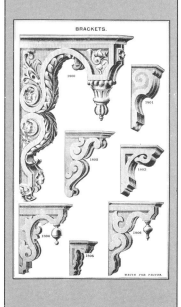

Machine-made, mail-order brackets from a 19th-century manufacturer's catalog.

Early coarse plaster mix of clay, lime and straw used as a base coat in the stairhall at Drayton Hall, 1738–42, Charleston, S.C. (NTHP)

linoleum moldings, offered by Sears, Roebuck and Company, were recommended to frame wallpaper panels or form cove, or concave, cornices. In the 1930s fiberboard products were introduced by such companies as Masonite. One type of board used for walls and wainscots had beveled edges, grooved designs and strips of metal covering the joints.

In many buildings constructed after World War I, specific historical references were no longer included and moldings were simplified. Some casings and baseboards were so generic that they were considered suitable for Colonial Revival, Craftsman or modern interiors.

PLASTER

The earliest plasters in American interiors used clay or lime as their basic cementing material, which was then mixed with water, a binder such as fiber or hair, and an aggregate such as sand. Clay was first used as the exposed wattle filling on frame houses and continued to be used as an interior finish at least through the 17th century. Clay also was the common interior and exterior surface coating on adobe buildings in the Southwest, where it is still used today. Lime, however, was usually the preferred ingredient in plaster from the 17th through the 19th centuries. Made from crushed limestone or shell deposits and often mixed with cattle hair, lime was used as early as the mid-17th century in Connecticut.

Before it was manufactured commercially, lime plaster was made on the building site in large pits lined with sand where quicklime and water were mixed together over a period of several days. For smaller jobs, quicklime was slaked, or mixed with water, in a wood or sheet metal-lined box. The lime paste could then be stored for long periods before being mixed with water, sand and fiber to make plaster.

GYPSUM PLASTER

Although gypsum plaster became popular in American interiors at the end of the 19th century, lime plaster continued to be used for finish coats because of its workability. Gypsum plaster, however, had the advantages of fast drying time and a harder finish. In the Middle Ages, quarries near

Paris supplied much of the gypsum needed for plaster — hence the term, "plaster of paris," now used to refer to all calcined gypsum, which is gypsum heated to remove water. Central New York State, where many water-powered mills for grinding plaster operated, became a major source for gypsum in the 19th century.

WALLBOARD

Wallboard made of gypsum between layers of paper was introduced in the late 19th century. It was favored for its easy installation and increased fire protection. An early type of gypsum board was "Sackett Board," which consisted of four layers of paper alternating with layers of plaster. Wallboard became so popular that by 1940 wall dimensions of new buildings were increasingly based on its modular sizes.

LATH

The earliest plasters were applied directly to inside wall surfaces or to a system of lath — thin pieces of wood that had been riven, or split with an ax — attached to the structural frame. Other types of wood lath included accordion lath — thin boards, split at opposite ends and pulled apart like an accordion when nailed to the framing — and sawn lath, introduced in the second quarter of the 19th century. The introduction of the circular saw permitted plasterers to create lath of uniform widths or thickness.

Toward the end of the 19th century metal lath came into use in the United States, although it had been patented in England as early as 1797. Because metal lath provides more spaces for plaster keys than wood lath and is resistant to rot and insect infestation, it was soon the preferred material and is still used by most contemporary plasterers. Modern types of metal lath include diamond mesh, expanded rib, wire mesh and sheet lath.

Board lath — mineral or vegetable products compressed between sheets of specially prepared paper — came into use around 1900 with the introduction of gypsum lath. Modern versions have been developed which are resistant to fire and high temperatures. A common type, rock lath, is made with surface paper containing gypsum crystals that bond

Riving wood lath. (Michael Devonshire)

with the plaster. Others include insulating lath, which provides a vapor barrier, and insulation board lath made of organic fibers or inorganic materials such as plastic or glass and used primarily on exterior walls and ceilings. Board lath resists fire, strengthens the framework structure and provides heat and sound insulation.

COLONIAL (1600–1720)

Along with wooden wainscots, lime or clay plaster was a common wall treatment in 17th-century colonial interiors. The earliest plasters were applied as a protective and insulating finish. They generally consisted of clay and hay (also known as daub or mud plaster) or lime and hair. The fireplace wall, usually an interior wall in houses with central chimneys, often had wood sheathing, while the inside faces of exterior walls sometimes were plastered. Lime plaster was favored for exterior masonry walls because of its insulation value, while wood sheathing allowed the transmission of heat from the chimney into the room.

Wood-frame buildings generally were constructed of heavy timbers with walls of studs and fillings such as daub or brick. For 17th-century English and Dutch wood-frame houses, plaster usually was applied over or between studs, either on lath or directly on the filling. Plaster and lath partitions, however, were not common until the 18th century because of the large number of nails, then made by hand, required to install wood lath.

The exposed beams, joists and bottom of the flooring of the story above served as a ceiling in most cases. Rarely, and only on the most expensive buildings, was plaster used for ceilings. The area between joists was sometimes plastered over lath, although ornamental plaster ceilings also might be constructed below the structure.

SPANISH COLONIAL (1600–1850)

Spanish colonial interiors of the 17th through mid-19th centuries generally were simple, with little ornamental detail. Although missions sometimes reflected the influence of the Spanish baroque style, the design of most secular struc-

tures was the result of folk building traditions. Adobe walls were plastered with mud or lime plaster and whitewashed. Mud plaster was applied with bare hands and polished with dampened sheepskin, deerskin or small rounded stones, creating a surface that undulated gently with the earth beneath it. In the early 20th century cement stucco began to be used as an interior and exterior finish on adobe buildings.

There was little in the way of ornament, although sometimes the last layer of plaster was tinted with ochre or pink pigments. Often a dado of darker adobe plaster and built-in adobe furniture such as benches or cupboards adorned a room. Ceilings generally were composed of vigas, or beams, and the exposed underside of the floor or roofing material above. In some interiors, beams were adzed to a rectangular shape and supported on corbels. In general, however, little or no wood was used in these buildings, because it was scarce in most of the areas settled by the Spanish and it also did not flex easily with the expansion and contraction of the adobe. Only more sophisticated houses, often built for Easterners, had molded baseboards, chair rails and picture rails, as well as mantelpieces with wood trim.

Left: Casa Amesti, Monterey, Calif., c. 1834, 1846. Bare white plaster walls, few, if any, wood moldings and an exposed ceiling structure were typical features of Spanish Colonial interiors, even in grand houses. (NTHP)

Right: Early 19th-century Spanish Colonial moldings from the Vallejo House, Sonoma, Calif. (Hannaford, *Spanish Colonial Buildings in California, 1800–1850*)

GEORGIAN (1700–80)

With the advent of the Georgian style in the early 18th century, plaster became more common as a wall and ceiling finish for buildings of all types and economic levels. However, elaborate wood paneling ornamented with the classical orders still was usually the choice for major public rooms. Plastered walls above paneled dadoes were another popular wall treatment at this time, as were plastered ceilings between encased ceiling beams and even flat plaster ceilings. As early as 1730, some buildings had cast plaster ceiling decorations.

FEDERAL (1780–1820)

Plaster more fully evolved as a construction material in the second half of the 18th century. New formulas for fibrous plaster (plaster reinforced with cloth, usually jute) that were developed in France and England made possible the casting of large sections of cornice and ceiling panels without cracks. The use of composition ornament (see Ornamental Plasterwork in this chapter) also allowed deli-

Elevation of the Federal-style dining room, Oak Hill, Peabody, Mass. The elliptical arch, delicate classical motifs such as urns, garlands and reeding and use of composition ornament on the door frame and arch were typical of high-style Federal interiors.

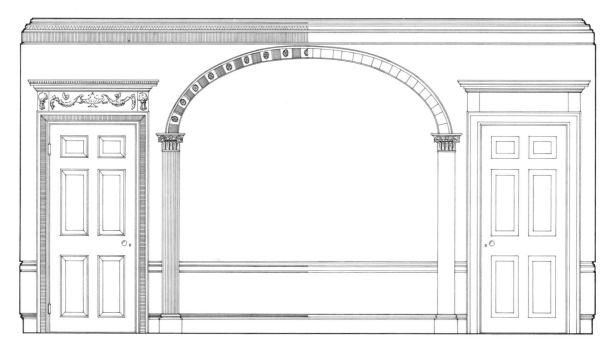

cate Adamesque, classically derived ornament for ceilings, cornice friezes, mantelpieces, architraves and wainscots — ornament that would have been prohibitively expensive if carved in wood or stone.

The most fashionable Federal interiors, including the banquet hall at Mount Vernon, made extensive use of this type of delicate plasterwork in the form of swags, garlands, flowers and bows. Ceilings often had fine plasterwork as well, usually comprising a central rosette of cast plaster, papier-mâché or wood, sometimes with matching fans or swag decorations at the border or in the cornice. Although simpler buildings generally had plain ceilings and less ornament, the decorative themes were similar.

GREEK REVIVAL (1820–60)

Plaster walls became increasingly popular as tastes turned toward the Greek Revival style. Ancient walls of stone or marble were better simulated with plain plaster than wood paneling. Room elevations reflected monumental Grecian proportions, classically symmetrical arrangements of doors and windows, and a new interest in Grecian motifs such as frets, honeysuckles and mythological figures. Ceilings of high-style rooms sometimes had cast plaster ornament in the centers, but most ceilings were left plain.

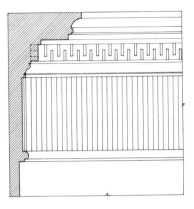

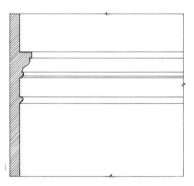

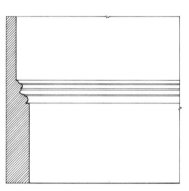

Above: Sections and partial elevations of a cornice (top), chair rail (center) and baseboard (bottom) at Oak Hill.

Left: Design for a plaster Greek Revival ceiling from Minard Lafever's 1833 guide.

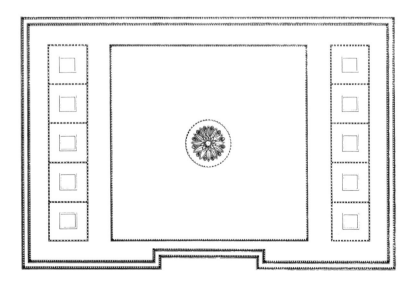

41

French Revival drawing room, John M. Davies House, New Haven, Conn., c. 1867–68, by Henry Austin and David R. Brown. (HABS)

Scagliola

Scagliola (imitation marble made from plaster of paris colored and mixed with pieces of marble, alabaster or granite and then highly polished) was popular for fine neoclassical interiors in the 19th and early 20th centuries. Another type of scagliola known as marezzo was colored using silk threads swirled through the plaster mixture before it set. Closely resembling marble in color, pattern, hardness and texture, scagliola is less costly than real marble and more durable than painted marbleizing (see Marbleizing in this chapter). It was used for major architectural elements such as columns and pilasters in important public buildings as well as fine houses.

VICTORIAN (1840–1900)

Plaster walls, ceilings and moldings continued to be popular for a range of styles throughout the 19th and early 20th centuries. Not only could plaster be molded into forms to suit each style, it also provided a smooth, seamless backdrop for wall treatments of almost any style.

Elaborate styles such as the Renaissance and French Revival styles exploited plaster's plasticity, adorning walls with ornamental plaster panels, floral friezes and bracketed cornices and creating coffered or paneled ceilings. Walls often were covered with wallpaper or new embossed products such as Lincrusta-Walton (see Other Finishes in this chapter). Several styles, moreover, exploited the textural possibilities of plaster. Italian Renaissance interiors around the turn of the century, for instance, had roughly textured plaster walls, usually with a narrow wood base and simple plaster or wood moldings.

SPANISH COLONIAL REVIVAL (1890–1930)

Spanish Colonial Revival interiors displayed coarsely textured plaster walls and ceilings, with a wavy, trowel-marked finish softened by bare hands. Areas around doors sometimes were richly ornamented with Spanish baroque decoration, but most decoration consisted of colored tile wainscots, friezes or borders. Spanish Colonial Revival interiors tended to be more stylized than Spanish Colonial interiors, evoking a mythic past that had little relation to Spanish Colonial architecture in America.

CRAFTSMAN (1895–1940)

Walls of Craftsman interiors often had a coarse or sand-textured finish, left natural gray or treated with a coat of tinted shellac or wax to create an overall appearance of handcrafted simplicity. Sometimes, though, wallboards were substituted in Craftsman interiors. Made of layered fiber, cardboard or wood pulp and first produced in the early 20th century, these were particularly useful for Mission and Craftsman designs where plaster walls were often divided into panels by wood battens.

PAINT

Until the mid-19th century, when wallpaper began to be mass-produced, plaster and wood walls often were finished with paint. Americans first used paint on exterior wood surfaces to protect against rot and warpage. By the late 18th century, as raw materials became cheaper and more easily obtainable, paint was used for decoration. Although at least one manufacturer was offering commercially mixed paints in the 18th century, painters often mixed their own paints and continued this practice through the third quarter of the 19th century. Packaged paints were introduced in 1858. By the 1870s ready-mixed paints were widely used. About this time paint companies published color cards with real chips of paint.

Before 1850 most paint colors were made from earth pigments or dyes such as ocher or indigo, generally imported from Europe. Purchased as dry powder or solid blocks, colors had to be ground and handmixed into the paint. By the 1870s aniline dyes made from coal tar were commercially available, after the discovery of the first such dye, mauve, in 1857. Aniline dyes allowed a greater and

Ornamental Plasterwork

Plasterwork traditionally was divided into two types: ornamental and flat. Ornamental plasterwork included molding made in special forms, or cast, and molding run in place at the site using a mold or template. Most cast ornament was enriched with surface decoration. Castings were mass-produced or cutom-made in wax or glue molds and were generally sold by the linear foot. Plasterers sometimes combined cast repetitive ornament with in-place molding of nonrepetitive designs. Wet plaster could also be stamped with small dies to create an ornamental effect.

Papier-mâché ornament from England and France was popular for mid-18th-century American interiors. Composition, or putty, ornament was also available. Made from powdered chalk, glue and linseed oil, pressed into molds and applied to surfaces when set, composition ornament allowed the use of finely detailed sculptural forms for a fraction of the cost of comparable wood or stone carving. It continues to be used today.

more consistent range of colors, but they faded quickly with sunlight and cleaning. In the 19th and early 20th centuries, paint mixtures used more stable white or red lead bases tinted with pigments. Now illegal in the United States, white lead has been replaced with titanium dioxide.

Whitewashed walls of a colonial interior contrasting sharply with extravagantly painted walls of the Victoria Mansion, Portland, Maine.

TYPES OF HISTORIC PAINTS

The base of all paint is one of three major binders: water, historically the least expensive; oil, introduced in the late 18th century; and casein, which lost favor around the mid-19th century.

Water-based paints

Whitewash was the least expensive and most popular paint from the 17th until the 20th century. Made from water and slaked lime, sometimes with the addition of salt, glue, sugar, rice flour, alum or oyster shells, it was used on walls and ceilings of all kinds of buildings, including adobe structures. While whitewash usually was applied to plaster, and woodwork was painted with oil paint (occasionally varnished or shellacked), utilitarian or modest buildings and less-used passages of most 19th-century houses often had whitewashed walls and woodwork. Whitewash was easy to make and apply, but it eventually flaked, and its coarse texture made it unsatisfactory for more formal interiors.

Distemper, or calcimine paint, closely related to whitewash, was used from the 18th century until the 1930s. Although not as durable as oil paint, distemper paint was favored because the materials — opaque watercolor paints made from calcium carbonate (chalk), tinting pigments, water and glue — were cheaper than oils, readily available, easy to work with, fast-drying and relatively odorless. Another major advantage over oil paint was that it could be applied immediately to new walls. New plaster had to season before walls could be painted with an oil paint, otherwise the plaster would absorb the oil as it dried, producing a mottled finish. Thus, distemper paint sometimes was used as a first coat until oil-based paint could be applied. Distemper paint was water soluble, however, and therefore not as durable or washable as oil paint.

Textured, or plastic, paint was popular between 1915 and 1935 for Craftsman and Colonial Revival buildings. Often used to imitate a tooled plaster finish, textured paint could produce a range of color and texture effects. It was packaged as a plastic powder with color, hot water and sometimes sand, to be added later. Plastic paint was applied with a wide brush and textured with a trowel, dry brush, sponge, wire brush or palette knife. It could also be finished to simulate ashlar or travertine.

Oil-based paints

Oil paints, composed of oil (usually boiled linseed oil), white or red lead and color pigments, were more expensive than their water-based counterparts but produced a much more durable surface. In mid- to late 19th-century buildings, it was a common practice to apply five coats of oil paint to plaster walls.

Casein-based paint

Casein paints used milk as a binder in a distemper or oil solution. Introduced in the early 19th century, casein paints were popular because they were more durable than distemper paints and cheaper than oil paints. They also dried rapidly and did not have a disagreeable odor. Unlike distemper paint, casein paint could not be dissolved by water and, thus, formed a relatively permanent coating.

Early recipes prescribed a mixture of slaked lime, skimmed milk, caraway, linseed or nut oil, and pigment. The demise of casein paint probably occurred with the mass production and increased affordability of oil paints in the late 19th century.

DECORATIVE PAINTING

Decorative painting became popular in America in the first half of the 18th century and continued into the early 20th century. Current styles and the availability of workers skilled in painting techniques influenced the prevalence of decorative painting. Until World War I most painters' skills included such decorative painting techniques as graining, marbleizing, stenciling, lacquering, japanning, gilding and often mural painting as well. Because many of these arti-

Decorative painting, Honolulu House, Marshall, Mich., showing a variety of painting techniques: marbleizing, stenciling, trompe l'oeil and freehand painting. (*The Old-House Journal*)

sans were itinerants and moved west with the settlers, most areas of the country display samples of their work.

Decorative painting served different ornamental purposes. In some cases, it was intended to imitate finer finishes than were actually affordable — expensive hardwoods, for instance, or wallpaper before it was mass-produced. In other cases, decorative painting became a sophisticated art of trickery or trompe l'oeil, camouflaging materials purely for the sake of novelty and demonstrating the cleverness of a designer or artisan. In these cases, the simulated product may have cost as much as the real thing.

Examples of decorative painting range from the parlor at Marmion (c. 1781) in King George County, Va., which had marbleized pilasters and painted scenes with flower-filled urns and leafy swags, to the house of tanner and tavern keeper John Jay French (c. 1850) in Beaumont, Tex., which had walls with false baseboards painted in bright blue and doors grained to resemble mahogany.

Decorative painting on a scale used at Lyndhurst (1838, 1864–65), in Tarrytown, N.Y., is almost unequaled. The dining room, added about 1865, is a collection of hidden secrets: what appear to be honey oak ceiling beams and corner cabinets are really painted pine; embossed leather walls are textured paint; wooden colonettes and baseboards have

Dining room, Lyndhurst, Tarrytown, N.Y., c. 1865. Here, decorative painting disguises the true nature of the original materials. (NTHP)

been marbleized to match the Italian marble mantel; and fire fenders of steel and wrought iron are finished to look like bronze. The entry hall's real marble floor is surrounded by walls painted to simulate marble blocks. In another passage, a plaster ceiling is grained to look like wooden boards with heavy Gothic molding.

Stenciling

Stenciling has been popular in America since the last quarter of the 18th century. It continued in favor through the Greek Revival period and then sporadically, as contemporary fashion dictated, into the early 20th century. The practice of stenciling may have been brought to America from England, but some sources suggest that its American origins were in areas of Pennsylvania settled by Germans from the Rhine Valley. Most of the existing stenciling dating from this period, however, has been documented in areas settled by the British.

In the late 18th and early 19th centuries, stenciling was often done by itinerant painters, each of whom had patterns handed down from generation to generation. Typically, a room may have had stenciled borders above the chair rail; around windows, doors and mantels; at the corners or tops of walls; or dividing the wall into panels. Continuous stenciling over entire wall surfaces in the manner of wallpaper was also done.

Motifs in the 18th and early 19th centuries were generally inspired by wallpaper, ornamental plaster, carved woodwork and pottery designs, simplified to suit the logistics of making and applying stencils. Designs included a range of geometric, natural and classical ornamental patterns: festoons and swags with tassels; bells and flowering vines; roses and leaves; and large-scaled decorations, particularly on overmantels, of birds, weeping willows, the federal eagle, and baskets and vases with flowers. Corners of plain walls were sometimes decorated with quarter fans, framed by narrow borders of geometric design or vines and flowers. Broad wall surfaces, often divided by stripes, displayed floral sprays, oak leaves, pineapples, stars, diamonds, sunbursts, hearts and similar designs. Often the same motif — used in various sizes in different places — decorated a single room.

Stenciling was also popular for Victorian interiors of

Traces of original stenciling under a later application of plaster and paint. (NTHP)

Turn-of-the-century Turkish-style room. Victorian styles such as this provided an excuse for covering every surface with texture and pattern. (Library of Congress)

many styles. The desired effect was the creation of patterned surfaces, and stenciling was often used in conjunction with wallpaper, wood paneling, leather or imitation leather, or tiles to create what was considered a unity of decorative elements. Flat, stylized designs with little attempt at naturalism were favored. Two-dimensional, nonrepresentational patterns were popular — geometric designs or highly conventionalized flowers, foliage and birds, generally using small-scale motifs spaced closely together for an overall effect. Pattern books such as Christopher Dresser's *Modern Ornamentation* (1886), *Studies in Design* (1876) and *The Art of Decorative Design* (1882–86) were important sources for stenciling motifs in the mid- to late 19th century. Owen Jones's *The Grammar of Ornament* (1856) also provided a source for historic and exotic motifs — Roman, Egyptian, Moorish and Pompeiian — that suited eclectic tastes.

The placement of stencils was more regulated than the style of motifs. Often wallpaper and stenciling were used together: a large-patterned paper on the walls with a stenciled frieze or ceiling. By the late 19th century, wallpaper was generally cheaper than stenciling an entire wall. Formal rooms, however, sometimes had complex stenciled designs over the whole wall. One popular motif was a diaper pattern of interlacing floral or geometric designs, frequently done in gold leaf or metallic paint.

Even when not used in conjunction with wallpaper, stencils were applied to particular areas of walls in Victorian buildings: the frieze, cove and "wipe line" — the area just above the dado or wainscot. Here a stencil pattern camouflaged the smear from dusting the top of the paneling. Ceilings also frequently had stenciled borders or lines or sometimes were divided into panels by decorative bands.

Stencilers of this period brought or cut their own stencils, frequently using published patterns. In the early 20th century stencils were sold in huge sheets for use on walls and ceilings. Sears, Roebuck and Company and Montgomery Ward stocked stencil patterns and tools, and instructional manuals, trade publications and illustrated sample patterns taught homeowners how to stencil. Patterns for corners, bands, dividers and allover patterns were available with motifs derived from historic and high-style sources.

In keeping with the handcrafted aesthetic, Craftsman interiors also favored stenciled friezes, often derived from Art Nouveau designs.

Graining

Graining and marbleizing, or "fancy painting," were used from the early 18th century until the early 20th century in all classes of public and private buildings. Their appeal was not just ornamental. They hid surface soil, when varnished were easier to clean than painted surfaces, and simulated more costly materials than many owners could afford.

In colonial interiors wood, particularly door and wall panels, was often grained to look like hardwoods such as

Historic interior with re-created wall stenciling and marbleized mantel. The different patterns denote the dado, field and frieze. (Jodi Monnich)

Left: Set of graining tools and templates advertised in an early 20th-century catalog.

Right: Wall design from Charles Eastlake's *Hints on Household Taste,* **1877. Eastlake recommended the tripartite division of the wall into dado, field and frieze by using different patterns or with wainscoting and moldings. (both, The Athenaeum of Philadelphia)**

cedar, mahogany, oak or maple. In the 19th century influential critics such as Andrew Jackson Downing recommended graining to imitate ash, maple, birch, oak or black walnut, claiming that a grained surface was the easiest one to care for and helped make a room look "furnished." In the first half of the 19th century, both grained and marbleized finishes were common on doors, window sash, baseboards, fireplace surrounds and mantels.

Realistic graining techniques were popular in Victorian buildings, although some critics, such as Calvert Vaux in his *Villas and Cottages* (1857) and Charles Eastlake in the 1870s, deplored the "sham and pretense" of graining. Others recommended hardwood paneling, at least for rooms such as entry halls and libraries. Nevertheless, most mid-century houses had painted or grained woodwork, usually pine, which was easily worked and inexpensive.

The reform styles of the late 19th and early 20th centuries, such as the Colonial Revival and Craftsman styles, deplored such dishonesty and advocated painted or naturally finished woodwork.

Marbleizing

Marbleizing was popular for neoclassical decoration in 18th-century France and England, ultimately becoming highly valued as trompe l'oeil. Marbleizing gained favor here in 18th-century interiors as increasingly affluent Americans began to appreciate decorative effects created by a variety of patterns and textures. Therefore, door frames and paneling, overmantels, stairs, baseboards (a pragmatic treatment, as they were easily soiled) and floors frequently were marbleized.

Floors over the years were variously marbleized, painted overall to imitate black-and-white marble squares or stenciled with a marbleized border around the edge. Marbleizing on door or wall panels often was done in conjunction with graining: panels were marbleized and stiles and rails were grained, or panels were grained and moldings were marbleized. In addition, styles such as the Greek Revival, which emulated classical architecture, used marbleizing on such distinctly classical elements as columns to heighten the antique effect.

Gilding

Traditionally used to simulate solid gold, gilding consists of the application of thin sheets of metal leaf to a prepared surface. Gold leaf, while very expensive, is the most permanent and durable of the metals, providing a soft, satiny luster to ornament.

An ancient Roman practice popularized in the late 18th century by Robert and James Adam, gilding was sometimes

Left: Marbleized columns. (Michael Devonshire)

Right: Gilded bracket from 1868. (HABS)

done in more sumptuous neoclassical American interiors of the early and late 19th century. It was particularly favored for rooms in the Greek Revival and Victorian Revival styles, including rococo, neo-Grec, French Revival and Renaissance Revival. Architectural elements such as column capitals, cornices, moldings, mantels and keystones sometimes were gilded, either with gold leaf or other, less expensive metal alloys such as Dutch metal. In interiors of the Gilded Age — between the Civil War and World War I — gilding served as an opulent status symbol for the wealthy. In the early 20th century gilded effects were also popular in commercial buildings such as movie theaters and hotels, as well as public institutions.

Now hammered by machine, gold leaf previously was produced by a hand process, rolled and beaten over and over again. It comes in leaves $3\frac{3}{8}$ inches square and $\frac{1}{250,000}$ of an inch thick. A book of gold leaf covers $1\frac{1}{2}$ square feet, while a pack covers 30 square feet. Also available is transfer, or patent, gold, which is leaf backed with sheets of tissue and applied with a gilding wheel. It is sometimes used in ribbons for architectural molding.

OTHER FINISHES

Although not the subject of this book, some mention should be made of other interior wall finishes that were popular alternatives to wood paneling, plaster and decorative painting. Authentic reproductions of some of these items, such as wallpaper, Lincrusta-Walton and Anaglypta, are available today for use in restorations.

WALLPAPER

Wallpaper was available in this country as early as the beginning of the 18th century. By the mid-18th century it was not uncommon in middle-class homes and by the 1840s was being manufactured here. With the advent of cheap mass-produced papers after the Civil War, wallpapers and borders around mantels, doors, windows, baseboards and chair rails became widespread in all sizes of dwellings.

Between the late 1860s and 1900, wallpaper was used almost ubiquitously on ceilings and walls, and often different patterns were used to represent a dado, field and frieze.

Other wall treatments of the late 19th and early 20th centuries included cartridge papers with designs printed on smooth stiff surfaces; ingrain, or oatmeal, papers, with color ingrained in the paper by pulp dyed before the paper was made; and "sanitary papers" printed in washable oil-based pigments.

Wallpaper used in conjunction with delicate Federal-style moldings in the James Ludlam House, Goshen, N.J. (Jack Boucher, HABS)

FABRICS

Decorative fabrics such as damask, silk, velvet, tapestry and chintz were also popular in wealthy post–Civil War American interiors, particularly those of a neoclassical or European taste. Before that, fabric was occasionally used in 18th-century houses imitating English interiors.

Another use of fabric wall covering was practiced in hot, humid Southern climates in the late 18th and early 19th centuries. Cheap cloth was sometimes laid on plaster walls before the scratch coat was dry to prevent the plaster from retaining moisture and rotting.

In the Southwest, 19th-century Anglo-Americans stretched cotton cloth over ceiling joists of adobe houses to protect rooms from debris falling from dirt roofs. The cloth was sometimes painted with a mixture of flour and water, causing the material to shrink and producing the effect of a plaster ceiling.

Fabric, particularly burlap, was also widespread in Craftsman interiors of the late 19th and early 20th centuries and was sometimes stenciled. Gustav Stickley, well-known editor of *The Craftsman* magazine, recommended wall panels covered with silk, canvas or Japanese grass cloth.

EMBOSSED MATERIALS

Other popular wall treatments of the late 19th century were such embossed materials as Lincrusta-Walton, Anaglypta, Tynecastle tapestry, Lingomur (a wood-fiber product), Japanese leather paper and papier-mâché repro-

Right: Appearance of embossed leather achieved with Lincrusta-Walton on an interior wall of the Molly Brown House, 1887–92, Denver. (National Park Service)

Below: Designs for a Lincrusta-Walton frieze and dado from a catalog of Frederick Beck & Company, c. 1884. (Winkler and Moss, *Victorian Interior Decoration*)

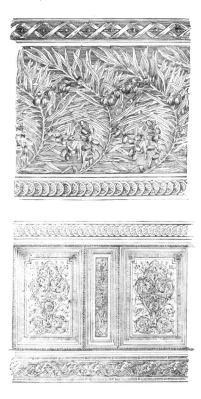

ductions of embossed and colored leather.

Lincrusta-Walton, which consisted of heavy canvas or waterproofed paper with embossed semiliquid linseed oil decoration, was invented in 1877 in England. It rapidly achieved widespread popularity, appreciated both for its rich ornamental effects and its durability. Applied much like wallpaper, it could be painted or highlighted to resemble wood, leather or metal.

Another embossed paper, Anaglypta, patented in England in 1887, was not as durable as Lincrusta but could be colored or glazed to suit any decor. Also fashionable, Tynecastle tapestry was canvas stiffened with glue and embossed with a pattern while still wet. Japanese leather paper simulated handtooled leather on embossed rolls of paper.

TIN

Between 1870 and 1930, walls and ceilings made of light metal sheets with stamped decorative designs simulating plaster motifs became popular. Most panels were nailed to furring strips, with concealed nails and joints. Although used in all types of buildings including residences, tin was particularly favored for commercial and institutional buildings. During the height of its popularity (1895–1915), more than 400 patterns were offered in a variety of styles: classical, Gothic, rococo and eventually Art Deco; brick, tile and stucco patterns also were available. Stamped-tin ornament came not only in ceiling and wall panels, but also in dadoes, chair rails, cornices, friezes and ceiling medallions and borders.

Stamped tin ceiling in the office of Thomas Cook and Sons, New York City, 1906. (Museum of the City of New York)

PRESERVING THE CHARACTER OF YOUR INTERIOR

Historic building interiors have been easy and frequent prey to changing fashions. Moldings could be replaced, walls recovered and wood trim repainted to provide a stylish and up-to-date image with far less effort than changing the roofline or altering window sash. The parlor of a late 18th-century townhouse could have assumed a series of identities in its first 150 years — Federal, Gothic Revival, Colonial Revival, to name a few. It may even have gone through a further transformation into a stripped-down modern interior, with almost wholesale removal of its historic ornament.

If you are lucky, problems with your old building are confined to cracked plaster or deteriorated varnish on wood paneling. More likely, however, the building will have undergone several remodelings. You may be confronted with missing or severely damaged decorative elements or a confusing array of alterations and additions. In those situations, you may want to hire a preservation professional to help devise a general approach to dealing with your historic structure before tackling the specifics of repairs. (See the chapter Planning for Rehabilitation.)

The first — and perhaps most critical — decision you will have to make is the choice of an appropriate treatment for the interior. This can range from restoring a structure to its exact appearance at a particular time to converting an interior to accommodate a new use. Rehabilitation, the strategy chosen by most owners of historic and other old buildings with special character, emphasizes preservation of existing historic features and materials, while making changes if necessary to support modern conveniences and uses, such as installing a new electrical or plumbing system but retaining visible features that define the historic character.

Moorish-inspired details, Willis Bristol House, New Haven, Conn., c. 1846. (Ned Goode, HABS)

Opposite: Victorian bedroom, Harriet Beecher Stowe House, Hartford, Conn. (Stowe-Day Foundation)

"Whatever you have in your rooms, think first of the walls, for they are that which makes your house a home."
— William Morris

Deciding whether to restore or to rehabilitate should be based in part on the historical or architectural significance of the building and the condition of its existing finishes and architectural elements. Is most of its historic detail intact? Does the building exemplify a particular architectural style or era? Was it designed by a well-known architect? Or was the building or its site associated with a person or event of particular historical significance? If the answer is yes to any of these questions, or if the building is already listed in the National Register of Historic Places or a local landmarks list, you may want to consider restoring both the exterior and interior. However, full-scale restoration sometimes is feasible only for buildings that will serve as museums. Even then, the cost of research into the building's history and appearance and the demands of scrupulously restoring or re-creating that appearance are beyond the means of many institutions and private homeowners.

On the other hand, if you are dealing with a structure of some architectural interest, but not of the highest significance, and it is intended for actual living or working space, you may want to consider a strategy of preservation and rehabilitation.

Below: Modest mid-19th-century interior. (National Park Service)

Right: Significant Georgian interior, Cliveden, 1763–67, Philadelphia. (NTHP)

While a precise restoration can be a painstaking and expensive process, it should not frighten anybody from bringing an old building back to life — respectfully. Far more old buildings demand treatment that falls into the category of rehabilitation. Factors such as a building's inherent significance, condition, planned use and cost often make rehabilitation the most feasible choice. This book focuses on rehabilitation as an overall preservation treatment. The advice concentrates on repairing existing interior features and finishes, although some aspects of restoration and re-creating missing elements are also covered.

BASIC APPROACHES TO REHABILITATION

Respect for the character and design of a building and its architectural elements and finishes should be the guiding principle in any rehabilitation project. You should try to retain as much of the original fabric or materials as possible, avoiding changing or replacing elements and respecting the original design when making necessary alterations and additions. The more historically or architecturally significant the building, the more careful you should be to preserve its historic features.

The most important thing to remember is that any change made in a building should be reversible. Old buildings belong to future as well as past generations and should be passed on with their history intact. Adding a gypsum board wall, for instance, is reversible. Removing an ornamental plaster ceiling is not. Neither is stripping woodwork that was originally grained and giving it a clear finish.

While you may need to make changes, you should always maintain records of what was there when the building came under your stewardship. As you proceed with the restoration or rehabilitation work, document whatever changes you make with before and after photographs and drawings. It also is important to leave evidence of the building's history intact, such as samples of all paint layers in inconspicuous places on all walls and moldings. In any case, the rule is always to alter the historic fabric as little as possible.

Modest interior with historic wood moldings and mantel. Partial removal of the baseboard to accommodate heating units should have been avoided. (T. J. Healey II)

59

At the heart of preservation philosophy is the goal of retaining and repairing historic elements and finishes. A three-pronged approach to rehabilitation is recommended by the National Park Service.

- Distinctive historic features should be *repaired*, not removed and replaced with all new material. Repairing, however, can involve the limited replacement of extensively deteriorated or missing parts of features. This is particularly suitable where there are surviving prototypes of repeated features such as units of an old tin ceiling or parts of an ornamental plaster ceiling medallion or cornice. The material used to repair the deteriorated parts can either be the original material or a compatible substitute material so long as it conveys the same visual appearance.

- If a distinctive feature must be totally *replaced* because it is too deteriorated to repair, its overall form and detailing should be used as documentation to create a matching replacement feature. If using the same kind of material for the replacement feature is not feasible from a technical or economic viewpoint, then a compatible substitute material can be used as long as it conveys the same visual appearance, that is, design, color and texture.

- If a distinctive historic feature such as a mantelpiece or stair is missing and insufficient physical evidence or other documentation exists on which to base a reconstruction, a replacement feature may be a *new design* that is compatible with the scale, design, materials, color and texture of the

Historic plaster ceiling medallion. The fine cracks in the ceiling can be repaired leaving the medallion in place. (National Park Service)

Reproduction ceiling medallions based on historic designs. (RestorationWorks, Inc.)

surviving interior features and finishes. Choosing a suitable replacement is a difficult decision. It is easy to create a false historical appearance by using over-the-counter reproductions or salvaged material from other buildings. You should seek advice from a preservation architect or your state historic preservation office.

SUBSTITUTE MATERIALS

The scarcity of craftspeople who can do custom plasterwork or create wood ornament is making the use of readymade reproductions more common. There is a long building tradition of adopting less costly and more convenient materials when necessary. Substitute materials are as old as architecture: plaster, for instance, was originally a replacement for stone. Some argue that composite polyurethane reproductions for cast plaster moldings follow in that tradition. While these reproductions have the advantage of being lightweight, flexible and relatively cheap, for the most part, they do not constitute historic preservation. If such a reproduction, however, matches a missing ornament in design, texture and overall appearance, it would be an acceptable substitute.

The ease and affordability of reproduction ornament make it tempting to over-ornament a building in a manner and style that may not be appropriate. If there is no historical basis for use of ornament, new ornament should not be installed. Custom woodwork, for example, may have been present in only one or two rooms originally. Replicating such woodwork for all rooms would be inappropriate.

SALVAGED MATERIALS

The same guidelines for using modern replacements in historic interiors apply to using salvaged historic elements from other buildings: if ornament such as wood paneling existed originally and is now missing or irreparably damaged, any replacement should match the original as closely as possible. In the absence of documentation as to an item's exact appearance, look at buildings of similar size, class, age and use in your area for ideas about the type and design of original ornament. Your use and choice of any salvaged materials must be carefully calibrated to your building's own history and appearance. The patina of age of salvaged items may persuade you they are appropriate for your old building, but if they were not there in the first place, they probably do not belong there now.

REHABILITATION STANDARDS AND GUIDELINES

The secretary of the interior, as head of the chief federal preservation agency, has codified a set of rehabilitation standards and guidelines for federal projects that are widely used by preservationists across the country. Initially established to evaluate work on properties listed in the National Register of Historic Places, the standards and guidelines are also used as a basis for certifying eligibility for the federal rehabilitation tax credit for income-producing properties.

THE STANDARDS

Whether your building is listed in the National Register or you intend to apply for a rehabilitation tax credit, the 10 Standards for Rehabilitation, updated in early 1990, provide sound principles regarding the treatment of any old or historic building. Although the rehabilitation standards suit many projects, the Interior Department has also developed overall standards for historic preservation projects that apply to other kinds of treatments — acquisition, stabilization, preservation, restoration, and reconstruction. These should also be considered where relevant.

The Secretary of the Interior's

Standards for Rehabilitation

1. A property shall be used for its historic purpose or be placed in a new use that requires minimal change to the defining characteristics of the building and its site and environment.

2. The historic character of a property shall be retained and preserved. The removal of historic materials or alteration of features and spaces that characterize a property shall be avoided.

3. Each property shall be recognized as a physical record of its time, place and use. Changes that create a false sense of historical development, such as adding conjectural features or architectural elements from other buildings, shall not be undertaken.

4. Most properties change over time; those changes that have acquired historic significance in their own right shall be retained and preserved.

5. Distinctive features, finishes and construction techniques or examples of craftsmanship that characterize a historic property shall be preserved.

6. Deteriorated historic features shall be repaired rather than replaced. Where the severity of deterioration requires replacement of a distinctive feature, the new feature shall match the old in design, color, texture and other visual qualities and, where possible, materials. Replacement of missing features shall be substantiated by documentary, physical or pictorial evidence.

7. Chemical or physical treatments, such as sandblasting, that cause damage to historic materials shall not be used. The surface cleaning of structures, if appropriate, shall be undertaken using the gentlest means possible.

8. Significant archeological resources affected by a project shall be protected and preserved. If such resources must be disturbed, mitigation measures shall be undertaken.

9. New additions, exterior alterations or related new construction shall not destroy historic materials that characterize the property. The new work shall be differentiated from the old and shall be compatible with the massing, size, scale and architectural features to protect the historic integrity of the property and its environment.

10. New additions and adjacent or related new construction shall be undertaken in such a manner that if removed in the future, the essential form and integrity of the historic property and its environment would be unimpaired.

Modest bedroom with Gothic Revival details. The historic character of this room is defined by the pointed arch window with flat board trim and the simple baseboard with shoe mold. The wallpaper also contributes to the room's special character and should be preserved if possible. (National Park Service)

To meet the basic rehabilitation standards, the Interior Department has developed accompanying guidelines for rehabilitating historic buildings and design and technical recommendations for the preservation and repair of old buildings. The guidelines are general rather than case-specific and advisory rather than absolute. Preservation treatments described in them should be tailored to specific circumstances in consultation with preservation professionals such as architects, conservators and architectural historians (see Professional Help in the chapter Planning for Rehabilitation) and after assessing the scope of the particular project.

The guidelines' priorities are, one, to ensure the preservation of a building's character-defining architectural materials and features and, two, to accommodate an efficient contemporary use. Beginning with the least intrusive actions, the guidelines provide recommendations for progressively more complex levels of work. These levels are:

■ Identification, retention and preservation of the form and detailing of architectural materials and features that are important in defining the historic character

■ Protection and maintenance of architectural materials and features

■ Repair of deteriorated architectural features

■ Replacement of severely damaged or missing architectural features

■ New additions to historic buildings

Included in the next pages are the rehabilitation guidelines that pertain to interior walls and ceilings, their finishes and decorative architectural elements. Approaches, treatments and techniques consistent with the Secretary of the Interior's Standards for Rehabilitation are listed in the Recommended column, while those that would adversely affect a building's historic character are given in the Not Recommended column.

Subdivision of an original lobby in a Victorian office building. Installation of a new wall has destroyed the proportions of the original space, while the paneling suffers from poorly located, surface-mounted conduit and electrical outlets. (Jonathan Sinaiko)

GUIDELINES FOR REHABILITATING

HISTORIC WALLS, CEILINGS, FINISHES

AND DECORATIVE ELEMENTS

Interior Spaces, Features and Finishes

An interior floor plan, the arrangement of spaces and built-in features and applied finishes may be individually or collectively important in defining the historic character of the building. Thus, their identification, retention, protection and repair should be given prime consideration in every rehabilitation project and caution exercised in pursuing any plan that would radically change character-defining spaces or obscure, damage or destroy interior features or finishes.

INTERIOR SPACES

Recommended	Not Recommended
Identifying, retaining and preserving a floor plan or interior spaces that are important in defining the overall historic character of the building. This includes the size, configuration, proportion and relationship of rooms and corridors; the relationship of features to spaces; and the spaces themselves such as lobbies, reception halls, entrance halls, double parlors, theaters, auditoriums and important industrial or commercial use spaces.	Radically changing a floor plan or interior spaces — including individual rooms — which are important in defining the overall historic character of the building so that, as a result, the character is diminished.

Altering the floor plan by demolishing principal walls and partitions to create a new appearance.

Altering or destroying interior spaces by inserting floors, cutting through floors, lowering ceilings or adding or removing walls. |

INTERIOR FEATURES AND FINISHES

Recommended	Not Recommended
Identifying, retaining and preserving interior features and finishes that are important in defining the overall historic character of the building, including columns, cornices, baseboards, fireplaces and mantels, paneling, light fixtures, hardware and flooring; and wallpaper, plaster, paint and finishes such as stenciling, marbleing and graining; and other decorative materials that accent interior features and provide color, texture and patterning to walls, floors and ceilings.	Removing or radically changing features and finishes which are important in defining the overall historic character of the building so that, as a result, the character is diminished.
	Installing new decorative material that obscures or damages character-defining interior features or finishes.
	Removing paint, plaster or other finishes from historically finished surfaces to create a new appearance (e.g., removing plaster to expose masonry surfaces such as brick walls or a chimneypiece).
	Applying paint, plaster or other finishes to surfaces that have been historically unfinished to create a new appearance.
	Stripping historically painted wood surfaces to bare wood, then applying clear finishes or stains to create a "natural look."
	Stripping paint to bare wood rather than repairing or reapplying grained or marbled finishes to features such as doors and paneling.
	Radically changing the type of finish or its color, such as

Unsympathetic alterations to a commercial Victorian interior, including false wood wainscot, flocked wallpaper and an original doorway filled in with bookcases. The pseudo-Victorian light fixture also is modern. (Jonathan Sinaiko)

Protecting a wood floor with plastic tarp before stripping woodwork. Edges are taped, plastic sheeting laid down and a heavy paper applied over the plastic and taped down. (Wayne Towle, Inc.)

Recommended	Not Recommended
	painting a previously varnished wood feature.
Protecting and maintaining masonry, wood and architectural metals which comprise interior features through appropriate surface treatments such as cleaning, rust removal, limited paint removal and reapplication of protective coating systems.	Failing to provide adequate protection to materials on a cyclical basis so that deterioration of interior features results.
Protecting interior features and finishes against arson and vandalism before project work begins, erecting protective fencing, boarding up windows and installing fire alarm systems that are keyed to local protection agencies.	Permitting entry into historic buildings through unsecured or broken windows and doors so that interior features and finishes are damaged by exposure to weather or through vandalism.
	Stripping interiors of features such as woodwork, doors, windows, light fixtures, copper piping and radiators; or of decorative materials.
Protecting interior features such as a staircase, mantel or decorative finishes and wall coverings against damage during project work by covering them with heavy canvas or plastic sheets.	Failing to provide proper protection of interior features and finishes during work so that they are gouged, scratched, dented or otherwise damaged.
Installing protective coverings in areas of heavy pedestrian traffic to protect historic features such as wall coverings, parquet flooring and paneling.	Failing to take new use patterns into consideration so that interior features and finishes are damaged.

Recommended	Not Recommended
Removing damaged or deteriorated paints and finishes to the next sound layer using the gentlest method possible, then repainting or refinishing using compatible paint or other coating systems.	Using destructive methods such as propane or butane torches or sandblasting to remove paint or other coatings. These methods can irreversibly damage historic materials.
Repainting with colors that are appropriate to the historic building.	Using new paint colors that are inappropriate to the historic building.
Limiting abrasive cleaning methods to certain industrial or warehouse buildings where the interior masonry or plaster features do not have distinguishing design, detailing, tooling or finishes; and where wood features are not finished, molded, beaded or worked by hand. Abrasive cleaning should only be considered after other, gentler methods have been proven ineffective.	Changing the texture and patina of character-defining features through sandblasting or use of other abrasive methods to remove paint, discoloration or plaster. This includes both exposed wood (including structural members) and masonry.
Evaluating the overall condition of materials to determine whether more than protection and maintenance are required, that is, if repairs to interior features and finishes will be necessary.	Failing to undertake adequate measures to assure the preservation of interior features and finishes.
Repairing interior features and finishes by reinforcing the historic materials. Repair will also generally include the limited replacement in kind — or with compatible substitute material — of those extensively	Replacing an entire interior feature such as a staircase, paneled wall, parquet floor or cornice; or finish such as a decorative wall covering or ceiling when repair of materials and limited replacement of such parts are appropriate.

69

Recommended	Not Recommended
deteriorated or missing parts of repeated features when there are surviving prototypes such as stairs, balustrades, wood paneling, columns, decorative wall coverings or ornamental tin or plaster ceilings.	Using a substitute material for the replacement part that does not convey the visual appearance of the surviving parts or portions of the interior feature or finish or that is physically or chemically incompatible.
Replacing in kind an entire interior feature or finish that is too deteriorated to repair — if the overall form and detailing are still evident — using the physical evidence to guide the new work. Examples could include wainscoting, a tin ceiling or interior stairs. If using the same kind of material is not technically or economically feasible, then a compatible substitute material may be considered.	Removing a character-defining feature or finish that is unrepairable and not replacing it; or replacing it with a new feature or finish that does not convey the same visual appearance.

MISSING HISTORIC FEATURES

The following work represents particularly complex technical or design aspects of rehabilitation projects and should be considered only after the present concerns above here have been addressed.

Recommended	Not Recommended
Designing and installing a new interior feature or finish if the historic feature or finish is completely missing. This could include missing partitions, stairs, elevators, lighting fixtures and wall coverings; or even entire rooms if all historic spaces, features and finishes	Creating a false historical appearance because the replaced feature is based on insufficient physical, historical and pictorial documentation or on information derived from another building.

Introducing a new interior feature or finish that is incompatible with the scale, |

Recommended	Not Recommended
are missing or have been destroyed by inappropriate "renovations." The design may be a restoration based on historical, pictorial and physical documentation; or a new design that is compatible with the historic character of the building, district or neighborhood.	design, materials, color and texture of the surviving interior features and finishes.
	Dividing rooms, lowering ceilings and damaging or obscuring character-defining features such as fireplaces, niches, stairways or alcoves, so that a new use can be accommodated in the building.
Reusing decorative material or features that have had to be removed during the rehabilitation work, including wall and baseboard trim, door molding, paneled doors and simple wainscoting; and relocating such material or features in areas appropriate to their historic placement.	Discarding historic material when it can be reused within the rehabilitation project or relocating it in historically inappropriate areas.
Installing permanent partitions in secondary spaces; removable partitions should be installed when the new use requires the subdivision of character-defining interior spaces.	Installing permanent partitions that damage or obscure character-defining spaces, features or finishes.
Creating an atrium or a light well to provide natural light when required for the new use in a manner that preserves character-defining interior spaces, features and finishes as well as the structural system.	Destroying character-defining interior spaces, features or finishes; or damaging the structural system in order to create an atrium or light well.

Original handmade wood lath preserved under new metal lath before replastering. (National Park Service)

Recommended	Not Recommended
Adding a new floor if required for the new use in a manner that preserves character-defining structural features and interior spaces, features and finishes.	Inserting a new floor within a building that alters or destroys the fenestration, radically changes a character-defining interior space or obscures, damages or destroys decorative detailing.

STRUCTURAL SYSTEM

Recommended	Not Recommended
Identifying, retaining and preserving structural systems — and individual features of systems — that are important in defining the overall historic character of the building, such as post-and-beam systems, trusses, summer beams, vigas, cast-iron columns, above-grade stone foundation walls and load-bearing brick or stone walls.	Removing, covering or radically changing features of structural systems which are important in defining the overall historic character of the building so that, as a result, the character is diminished.
	Demolishing a load-bearing masonry wall that could be augmented and retained and replacing it with a new wall (e.g., brick or stone), using the historic masonry only as an exterior veneer.
	Leaving known structural problems untreated, such as deflection of beams, cracking and bowing of walls or racking of structural members.
Examining and evaluating the physical condition of the structural system and its individual features using nondestructive techniques such as X-ray photography.	Utilizing destructive probing techniques that will damage or destroy structural material.

Recommended	Not Recommended
Replacing in kind — or with substitute material — those portions or features of the structural system that are either extensively deteriorated or are missing when there are surviving prototypes such as cast-iron columns, roof rafters or trusses, or sections of load-bearing walls. Substitute material should convey the same form, design and overall visual appearance as the historic feature; and, at a minimum, be equal to its load-bearing capabilities.	Installing a replacement feature that does not convey the same visual appearance, e.g., replacing an exposed wood summer beam with a steel beam.

ENERGY RETROFITTING

Recommended	Not Recommended
Installing insulating material on the inside of masonry walls to increase energy efficiency where there is no character-defining interior molding around the window or other interior architectural detailing.	Applying urea formaldehyde foam or any other thermal insulation with a water content into wall cavities in an attempt to reduce energy consumption.

Resurfacing historic building materials with more energy-efficient but incompatible materials, such as covering historic masonry with exterior insulation. |
| Retaining historic interior shutters and transoms for their inherent energy-conserving features. | Removing historic interior features which play a secondary energy-conserving role. |

PLANNING FOR REHABILITATION

To plan a rehabilitation strategy, you need to know as much about the history of your building as possible. The existing structure may be the product of a number of generations of construction, with rooms and stories added as its occupants' needs changed. The story of one house illustrates the complexity you may be facing: beginning as a one-room shanty in 1820, it was enlarged to a 1½-story cottage in 1825 with a built-in cupboard, fireplace and several windows. Between 1830 and 1835 the cottage was turned into a summer kitchen; at that time, room partitions and a gable were added. In 1875 the interior was remodeled with decorative plasterwork and once again, in 1920, with tongue-and-groove paneling and tin ceilings.

How do you determine what happened when and how do you decide what to preserve? The best advice is usually to preserve the historic character of what is there — the cumulative history of a structure and its occupants. Unless a building is exemplary of a particular style or is associated with a particular historic event, it should not be restored to a specific period if that means removing or demolishing later additions and remodelings that contribute to the historic character of the building. The current philosophy of preservationists is that a building is the sum of its history: early as well as later alterations may be equally worthy of preservation.

For example, the 1920s tongue-and-groove paneling and tin ceilings of the house described earlier should not be removed in order to restore or re-create the ornamental plasterwork of its 1875 appearance, just as the attic story, partitions and fireplace should not be removed to re-create the original shanty. On the other hand, if recent nonsignifi-

> "One must always remember — in all facets of architectural preservation — the house is the master."
>
> — Peter John Stokes, Restoration Architect

Fireplace at the Ansley Wilcox House, Buffalo, N.Y., and stairhall, Devereaux House (opposite), 1857, Salt Lake City, both before restoration. (HABS)

cant alterations such as Masonite paneling, a dropped ceiling, vinyl flooring over a hardwood floor or inappropriate moldings or finishes detract from the overall historic character of the interior, these should be removed and the historic finishes or molding restored or re-created. In summary, the treatment of rehabilitation should respect *all* historic materials and features.

RESEARCHING THE BUILDING'S HISTORY

The more you know about your building, the better off you will be in planning its future. A combination of archival and architectural evidence may be necessary to determine its full history. You can do much of the archival research yourself. However, if a building's history is complex or if extensive rehabilitation work is likely, you may need a preservation architect or architectural historian to help you sort through the records and match them with the building itself.

Try to learn the approximate date and style of the building and all its significant additions and remodelings. It is also helpful to know who designed and constructed the building, as well as the status of the original client and subsequent users or inhabitants. A building custom-designed by an architect for a particular client probably would have had a different level and quality of finishes from a building constructed by a speculative builder.

ARCHIVAL RESEARCH

Assemble all available written documentation, photographs and drawings that describe the evolution and appearance of your building over time. Good sources for background historical information on old buildings are historic site surveys, which have been done in many communities and should be available at your local or state historic preservation office or library. These generally describe the major architectural features and the history of construction and ownership of a building and may include a bibliography.

Public buildings such as town halls, courthouses, hospitals and government office buildings were usually well doc-

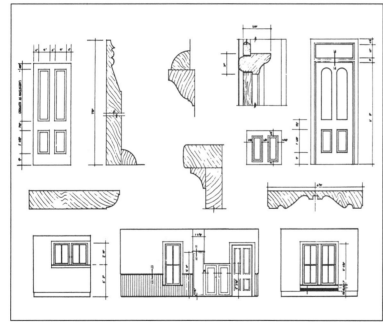

Top left: Dining room looking west, Eisenhower House, Abilene, Kans. (David von Riesen, HABS)

Left: Elevations from the Eisenhower House dining room. Measured drawings such as these are used to record historic features and help plan the incorporation of any required new elements in ways that will be the most sympathetic to existing historic fabric. (Thomas B. Simmons, HABS)

umented in building committee records. This information should be stored in city or state archives. The construction of churches was often chronicled by parish members, while commercial buildings and houses of important or wealthy members of the community were frequently written up in local newspapers. Other sources of information or photo-

Historic photo of the dining room in Terrace Hill, Des Moines, Iowa. Photos such as this can provide clues to historic features and treatments for appropriate rehabilitation or re-creation, if elements are missing or damaged beyond repair. (National Park Service)

graphs are libraries, historical societies, title guarantee companies, and wills, deeds and tax records kept in local courthouses.

ARCHITECTURAL RESEARCH

Once available written and visual information has been collected, the building itself will provide further clues to its history. Do not try to date interiors by the age or exterior style of a building, however. Remember that exterior and interior styles often did not match, and rooms within the same building did not necessarily synchronize with each other. Not only were particular styles favored for certain rooms, no matter what the exterior style, but interiors also were frequently remodeled to keep pace with changing tastes. The date of interior finishes and decorative elements may be much later than the building's construction. In that case, an interior is best rehabilitated as it is, preserving and repairing its existing historic features, rather than restoring it to a conjectured original appearance.

Structural changes such as additions and alterations often can be detected in the basement or attic where the frame is exposed. Different sizes, methods of fabrication and

Basement of a c. 1920 row house in Baltimore. Additions and alterations may sometimes be apparent in the exposed structural framing of basements. (The Peale Museum)

means of fastening the framing members may indicate different construction dates. A local preservation architect or architectural historian will be helpful in interpreting these clues. The trained eye of someone familiar with the chronology of local building techniques can distinguish ax-hewn beams from pit-sawn and circular-sawn beams or hand-wrought from machine-made nails. They will be able to put together the evidence to establish approximate dates for alterations.

In finished rooms seams in floorboards may indicate the former locations of walls. Clues such as changes in molding and paneling styles also may indicate remodeling, and their profiles and methods of fabrication can sometimes be dated (see the chapter American Interior Styles). Different finishes were popular at different times, and these, too, may be dated. Again, an architectural historian or preservation architect can help decipher these clues (see Professional Help in this chapter).

There are other questions you can ask during the planning stage. When was the tongue-and-groove paneling installed? What type of wood is it? Was it originally painted or did it have a clear finish? Was it custom-made or mass-produced? Answers to questions like these will help you determine how to rehabilitate your interior, to select both specific treatments for particular elements as well as the scale and type of finish for the remainder of the room or replacement elements if they are required. Later chapters in the book offer general guidelines for evaluating wood, plaster and painted walls and planning their rehabilitation.

If the interior of your old building has been stripped of most of its historic molding and wall finishes, what can you do? Re-creation of an earlier appearance should be based to the greatest extent possible on documentation of what actually was there. If available, original drawings or photographs may provide some clues. Pieces of molding left in the building (check inside closets if the rooms yield nothing), original wall and ceiling finishes hidden behind dropped ceilings, paneling or pipe chases installed after the building was constructed — you may be surprised by what a little detective work will uncover. Surviving fragments of the historic fabric should then become the basis for restoration or re-creation of moldings and finishes.

If there is little or no documentary or architectural evi-

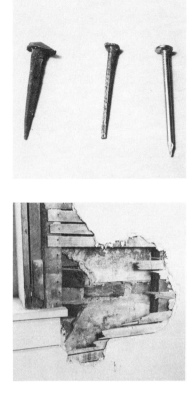

Top: Lath nails from the 18th, 19th and 20th centuries. Types of nails can help date architectural features. (Michael Devonshire)

Above: Custom House, c. 1760, Chestertown, Md. A cross section of this building's history is revealed here: original plaster was applied directly to the exterior masonry wall and a chair rail, now missing, had been installed; later, the masonry was furred out and machine-sawn wood lath nailed down before an application of three-coat plaster. (Maryland Historical Trust)

dence of either the original or a later appearance of a room, your choices are both broader and more difficult. You should seek guidance from a preservation professional, who may advise you to take stylistic cues from the building's exterior or to look at manuals of decoration popular at the time of the building's construction that were used by local architects and builders in your area. It is important, however, to consider the appropriateness of any proposed new treatment for your particular building. Study interiors of buildings of the same type, style, date and class in your area to get an idea of the relative scale and proportion of architectural elements and finishes. You also may elect to have a contemporary room, on the other hand. In the absence of existing historic material or documentation, this is a valid option.

Evidence of the past in the Hackley House, Muskegon, Mich. Removal of a modern sink and cabinets reveals the location of an earlier coal hoist enclosure (light panel in floor). Marks on the wainscoting to the left indicate the location, size and style of the original sink. (National Park Service)

SURVEYING EXISTING CONDITIONS

Before beginning rehabilitation, existing conditions should be documented as part of a comprehensive survey of work to be done. This survey will help you plan the project so that work can be performed in a logical and orderly manner.

While this book focuses on the finishes of interior spaces, it is critical to first determine the state of the structural, electrical, plumbing and mechanical systems behind the wall and ceiling surfaces. The presence of asbestos should be investigated at the same time. The condition and adequacy of these building systems is best analyzed by an architect or engineer, who can also provide specifications for their repair and any required upgrading. A preservation architect can plan how to incorporate new systems or repairs into the building with the least amount of disturbance to historic features and materials.

BUILDING SYSTEMS

Upgrading the electrical, plumbing, heating and ventilating systems can wreak havoc on the interior of your old building if not planned carefully by someone sensitive to preservation. The following methods of accommodating repairs and alterations with minimum damage to historic interiors should give you an idea of general strategies, but any treatment must be tailored to your particular building, its use and local building code requirements.

Electrical systems

Repairing or upgrading the electrical service requires work inside walls. But nobody expects you to live in a building with faulty wiring or an insufficient number of circuits or outlets for the sake of preserving the plaster. Wholesale destruction of historic plaster and lath is usually not required for rewiring a building — although too frequently the strategy of insensitive contractors. If you consider the cost of repairing damaged walls, less destructive methods make financial as well as preservation sense.

How can upgrading be done with sensitivity to historic materials? One solution is to take advantage of other work being done behind a wall, such as plumbing or structural repairs, and to install electrical wires while the wall is open. Also, wires can frequently be routed through unseen areas that are readily accessible: pipe chases, unused vents, crawl spaces, closets, the basement ceiling, between joists of the floor above. Baseboards, door trim and sometimes even cornice molding also can be removed to provide channels for conduits behind the wall. Another technique, called "fishing," involves cutting a small hole at one end of a wall or ceiling and sending the wire through to the other end. Studs, fire stops or lumps of plaster may obstruct the wire's path, however. Even surface-mounted conduits, if located discreetly — along the edges of baseboards or molding, for instance — are preferable to demolishing historic plaster or other finishes.

Plumbing systems

Similarly, most of a plumbing system is located behind walls, so that repairs or upgrading can result in disaster to your historic interior. Careful planning, however, will mitigate the damage. As with new wiring, hidden crawl spaces, abandoned vents and closets can serve as locations for new pipe chases. Or if a new bathroom fixture is being installed in an upper story, see if the trap can be located between, instead of underneath, floor joists, where it would require lowering the ceiling below. Also, supply and waste pipes

Insensitive mounting of modern conduit and light fixture on a wood beam. Careless smears of paint mar the wood's clear finish. (Jonathan Sinaiko)

can often be run through built-in furniture such as cabinets and vanities.

Heating and ventilating systems

Most old buildings have excellent built-in thermal properties. Before cheap energy allowed environmentally insensitive planning and design, buildings were sited, designed and landscaped to capitalize on natural sources of heating and cooling. The number and size of windows were planned carefully on each building face to take advantage of or to minimize the effect of sun or wind on the interior. Shutters, blinds and awnings also were used to mediate heat gain and loss through window openings. In warm climates such features as overhanging roofs, wide eaves and porches protected buildings from sun-caused heat gain. Landscaping also was used strategically; for instance, deciduous trees often were placed to shield buildings from summer sun while letting in winter sun.

Air conditioning. Before resorting to air conditioners or central air conditioning, which can have severe disfiguring effects on your old building, try some of these historically proven measures:

- Awnings
- Shutters
- Attic fan
- Ventilators in eaves
- Shades
- Landscaping
- Ceiling fan

If, after trying more passive measures, you decide to install central air conditioning, try to have the ducts run in behind-the-scenes spaces such as closets and to locate registers in discreet places that do minimal damage.

Heat. Before expanding the heating system or insulating walls, it is wise to see what less drastic measures can achieve:

- Caulking holes and cracks
- Installing weather stripping
- Insulating the roof, basement and crawl space
- Using heavy curtains with a hood or pelmet, shutters, internal storm windows, clear plastic sheets or opaque thermal insulated panels on windows
- Sealing seams and insulating heating and cooling pipes, ducts and the hot water heater

Frame house in Indianapolis. Features such as louvers in the gable, shutters, deep porches and shade trees were historically used to help cool buildings naturally. (NTHP)

■ Maintaining heating equipment for maximum efficiency

If upgrading a heating system is necessary, it is best to extend the existing system by adding dampers, registers, radiators and similar equipment. Installing an entirely new system could result in significant loss of historic features and materials, including the mechanical system itself.

Insulating exterior walls can be very expensive and entail a loss of historic material as well. It can require large-scale demolition of exterior or interior walls and also cause far-reaching damage if insulation is not installed and vented properly or if the wrong type is used. If insulation is determined necessary even after all the other energy-saving methods have been explored, careful research should be done into the appropriate type and methods of installation. In wood-frame buildings batt insulation can be installed in the wall cavity by carefully removing and reinstalling the exterior siding. If the siding cannot be removed without serious damage, blown-in or injected insulation is another possibility. The wrong type of insulation, however, can cause severe deterioration in historic building materials. Buildings with masonry cavity walls usually have good thermal properties already, and insulation applied in the cavity can cause severe moisture problems. Insulation can be applied to the interior face of the walls by furring out the wall, but the resulting damage makes this a choice appropriate only for interior spaces with little or no significance.

Solid-panel shutters in a colonial interior in Norwich, Conn. (above, NTHP), contrasting with louvered Victorian shutters in Georgetown, Colo. (top, The Georgetown Society).

ARCHITECTURAL ELEMENTS AND FINISHES

Evaluation of a building's infrastructure should be accompanied by a survey of architectural elements such as walls, ceilings, floors, windows, doors, fireplaces and their finishes and ornamentation. Hiring a preservation architect is your best bet for this work. Sometimes problems apparent in the finishes can suggest troubles of a more fundamental nature that can be diagnosed only by someone with a trained eye. Crumbling plaster, for instance, may be the result of dampness caused by a leaking roof or poor drainage system, faulty plumbing, inadequate ventilation or seepage around the foundation. Or the pattern of cracks in a plaster wall may indicate problems of a structural nature. (See individual chapters on wood, plaster, and painted finishes for

Top: Water-damaged plaster possibly indicating leaks that should be corrected before beginning plaster repairs. (Michael Devonshire)

Above: Wall area indicating damage to plaster, degraded finish on wood paneling and a poor wood patch. (Jonathan Sinaiko)

further discussion of diagnosing problems.) In any case, troubles behind floor, wall and ceiling surfaces must be corrected before beginning to rehabilitate the finishes. Too often access to structural and plumbing systems seems to be through freshly patched and painted walls and ceilings!

The initial survey should note all significant architectural elements, their materials and existing condition, as well as whether they appear to be original or the result of historic or modern alterations. Remember that alterations are not necessarily bad, as they are evidence of a building's evolution over time. Only after a survey of an entire building has been made can the merit of alterations be evaluated.

Areas of proposed rehabilitation should be documented in their current conditions with photographs and drawings noting critical dimensions. Even if the work will focus on small-scale details, general interior views should be photographed — including rooms that seem to be unremarkable. It is also important to keep a log of the photographs, noting the negative number, subject, location, camera position and the date, for future reference and before-and-after comparisons. As work proceeds, it is a good idea to continue to keep a written and photographic log. This provides a record for future maintenance and repairs and helps you as well as future generations keep track of your work.

PROFESSIONAL HELP

Depending on the scope of rehabilitation, you may need a preservation consultant such as an architect, engineer, architectural conservator or historian to work with you. Not all architects and engineers have experience in the proper rehabilitation of old buildings, so do not pick names randomly out of the Yellow Pages. Check with your local or state historic preservation office, historical society or American Institute of Architects chapter for recommendations. It is a good idea to meet with several professionals to discuss the project before committing yourself. Talk to their previous clients and choose the consultant who seems most knowledgeable about the history and rehabilitation of old buildings and understands your objectives as well as your budget.

Cosmetic improvements such as plaster repair or wood

Patching crack in plaster wall. Some simple repairs such as this can be handled by the homeowner, but a professional should be called in where larger areas need to be replastered. (Jack Boucher, HABS)

refinishing can usually be done without a permit, but you should check with your local building department to determine legal requirements. You may need the services of an architect or engineer if a building permit is required. Drawings for the proposed work will have to be made and stamped by a licensed architect or structural engineer. (Note: Building codes set minimum standards for construction safety, but do not guarantee quality of work.)

ARCHITECTS

Whether or not a permit is required, an architect familiar with local resources and building traditions can help correlate archival research with architectural evidence to date features of a building. He or she also will be able to evaluate existing conditions and determine necessary repairs, as well as research applicable building codes for requirements that may affect the work. An architect will do the initial documentation drawings and the working drawings and specifications detailing rehabilitation construction methods and materials. The working drawings and specifications make up the contract documents, which then become part of the contract between the owner and contractor.

Fees

Architects work on a fee basis, generally a percentage of the construction cost, or on a time-and-materials basis, usually with limits established for each phase of work. They may in turn hire consultants for specific aspects of the job such as original paint color research or new electrical work.

Contracts

A contract between the architect and owner should be drawn up detailing the following:

■ Responsibilities of the owner and architect during each phase of work: schematic design, design development, construction documents and construction

■ Any additional services beyond the basic services, including the payment of consultants' fees

■ Schedule for the performance of the work

■ Fee basis and list of reimbursable expenses

■ Schedule of payments

Bids

Once the drawings and specifications have been completed for the rehabilitation, they are sent out for bid to several contractors. Make sure the bid includes a cost breakdown, so that you and your architect can evaluate and modify different items, if necessary. Bids are nonbinding until they are incorporated into a signed contract.

The architect will help evaluate the bids, and a contract is then negotiated between the owner and contractor. The contractor may in turn hire subcontractors to perform specialized aspects of the work, for example, ornamental plaster repair. The architect serves as the owner's representative and will monitor the quality of the work to ensure that it conforms to the drawings and specifications.

CONTRACTORS

If the scope of work involves several trades, you should consider hiring a general contractor to supervise the job. It may be tempting to try to save money by acting as your own general contractor and hiring and supervising the various subcontractors yourself. If your time and experience are limited, however, this can ultimately end up costing you more in terms of hassles and coordination-related delays.

Contractors' bonds and insurance

Licensed contractors are bonded and must do the work as prescribed by local codes. There are different types of bonds, and you should check which ones your contractor has:

Contractor's license bonds, usually for a minimum of $5,000, are the only bonds required of licensed contractors. They are not guarantees of performance or financial responsibility.

Performance bonds ensure completion of the work according to the plans and specifications. The bonding company may settle for damages or put another contractor on the job if the first one does not complete it satisfactorily.

Payment bonds protect owners from liens for materials or labor against their property.

Contract bonds guarantee job completion and payment of all materials.

Ask your contractor to furnish a completion or contract bond as well as the license bond. And make sure the bond amount covers the cost of the job.

A licensed contractor must also carry worker's compensation insurance and will probably have property damage and liability insurance as well. If the contractor is not insured, you may be liable for injuries or damage to public property. Have your lawyer draft a liability waiver if the contractor's insurance is not adequate.

Fees

Contractors charge either a fixed fee for their work or a fee based on time and materials plus a certain percentage for overhead and profit. Although you may ultimately save money on a time-and-materials contract if the job is done in a shorter time than the contractor estimated, most people prefer to have a fixed final cost. Even this may be modified by change orders, however, if the owner or architect alters the scope of work or materials from the original contract. Make sure that the schedule of payments is based on the contractor's performance and amount of work completed and that an amount (usually 10 percent of the cost) is retained until the job is completed.

Contracts

Whether or not you use a licensed contractor, no work should be performed without a written contract. Check your own insurance policy and mortgage for clauses regarding remodeling work. The contract should at least describe the following:

- Name, address and license number of the contractor
- Scope of work (plans and specifications, which describe the work to be performed as well as the type and quality of materials, should be included as part of the contract)
- Statement requiring that all work conform to federal and local codes
- Schedule of work, with approximate start and finish dates and including a penalty or liquidated damages (usually calculated on a per-diem basis) if completion is delayed
- Total price of the work with a breakdown of costs, as well as a description of unit costs for change orders
- Schedule of payments
- Penalty for cancellation of the work
- Contractor's responsibility for obtaining permits (if you obtain the permits, you will be liable if the work is not done to code standards)
- Contractor's responsibility for clean-up and debris removal
- Contractor's bonds and insurance
- Respective liabilities of contractor and owner

It is important to include everything you expect the contractor to do in the contract. Otherwise, it probably will not get done.

ARTISANS AND CONSERVATORS

For restoration work — of original decorative painted finishes, for instance — professional conservators are likely to be needed. Names of specialists and artisans may be available from your local or state historic preservation office, a craft guild or the American Institute of Conservators. It is also a good idea to obtain recommendations from those who have had work done on a similar scale as your own. A craftsperson who regularly refinishes woodwork on individual houses may not be equipped to manage large-scale jobs, for which commercial methods may be suitable. For craftspeople as well as contractors, it is important to see examples of their work and to speak with their clients.

Craftsperson demonstrating how window sashes are made using a variety of hand planes. (Steve Consiglio)

REHABILITATING WOODWORK

Woodwork may be afflicted with a range of problems, from finish that has discolored with age to molding and paneling that are missing entire pieces. In between those two extremes may be damage such as blistered, crazed or checked finish; heavy buildup of finish obscuring the wood's character or detail; white haze or "blushing"; stains; warpage; loose panels; peeling or loose veneer; cracks; holes; gouges; breaks and rotted areas.

Most old woodwork suffers from at least one of these problems, and some may suffer from all of them. Age, heat, humidity, dryness, moisture, sun, insects — the causes of damage are many. Unless your woodwork has been pre-served in a bell jar for most of its life, it undoubtedly shows signs of environmental as well as human-inflicted abuse. Fortunately, most damage can be fixed by a craftsperson skilled in wood repairs and refinishing.

WOODS AND FINISHES

The more you know about your woodwork, the more informed your rehabilitation strategy will be. Particular types of wood in combination with specific finishes were popular at different times and for different styles. Knowledge of the type of wood used for your trim, for instance, may give clues about the original builder's intentions. Expensive or distinctive woods may have been exposed and enhanced with a clear, or natural, finish, while less distinctive woods may have been painted or camouflaged with graining or stains.

It is also helpful to know the type of wood, as different species may require particular stripping and refinishing

Sections and elevations of machine-made corner blocks from *The Victorian Design Book*, 1904.

Opposite: Test cleaning paneled wood ceiling at the Bradbury Building, Los Angeles. (Lynda Lafever; Jonathan Sinaiko)

Interior, Ruthmere, Elkhart, Ind., restored to c. 1910. Highly figured mahogany wainscoting and elaborate moldings contribute to the richly patterned effect of this late Victorian interior. (Andrew Hubble Beardsley Foundation)

techniques. Also, if any of your woodwork is missing or damaged beyond repair, replacement pieces ideally should be of the same species.

HISTORIC WOODS

All woods for buildings are classified in two categories: hardwood or softwood.

Hardwoods are cut from broadleaf (deciduous) trees and include species such as oak, walnut, mahogany and maple.

Softwoods come from coniferous trees such as pine, spruce and redwood.

Generally, hardwoods are heavier, harder, contain less resin and have a less uniform fiber structure than softwoods. The names may be misleading, however, as some softwoods are harder than some hardwoods.

A number of hardwoods, including oak, walnut and mahogany, have large pores in their cellular structure and are called open-grain woods. They sometimes require special procedures for finishing. Woods also vary in weight, strength, workability, color, texture and grain pattern. These characteristics determine their suitability for different purposes (structural or ornamental) as well as for particular finishes.

Bush-Holley House, Cos Cob, Conn. The 18th-century pine paneling was painted and given a rough graining effect. (Jack Boucher, HABS)

Common woods used in historic interiors for finish woodwork included ash, basswood, birch, cedar, cherry, mahogany, maple, oak, pine, rosewood, walnut and white-wood (poplar). Pine was by far the most popular wood for trim because it was cheap, plentiful and easy to work. When owners could afford it, however, pine was often painted, stained or grained to give it more distinction.

The more expensive and distinctive woods were generally used in public rooms, and the cheaper grades were used in private or "back-of-the-house" rooms. However, this practice varied with the type or cost of the building, as well as the taste of the times. In Greek Revival buildings, for example, painted (also grained or marbleized) woodwork was generally popular throughout interiors, although fine buildings sometimes had high-gloss clear finishes on expensive woods such as mahogany.

HISTORIC FINISHES

The purposes of finishing wood include protecting it from decay, insect infestation and moisture; producing a hard, flexible and durable coating resistant to damage and wear; and, in the case of clear finishes, enhancing the wood's texture and color.

Doors, trim, molding and paneling usually were finished in one of three ways:

- Clear or natural finish, exposing and enhancing the wood grain
- Grained, with a painted finish simulating wood grain
- Painted

(See the chapter Reviving Decorative Painted Finishes for discussions of grained and painted finishes.)

Softwoods such as pine were generally intended to be painted and should remain painted if that was the original treatment. High-quality hardwoods such as mahogany were usually given a clear finish, but there were no absolute rules. Hardwoods such as oak, for instance, were highly prized in Craftsman-style buildings and their natural characteristics enhanced with clear finishes. In other eras, however, oak was disguised by paint, stains or graining.

HISTORIC STAINS

Dyes and stains have long been popular historically to enhance or disguise woods. In the 18th and early 19th centuries, more than 100 different staining materials were used, including Brazilwood, logwood, archil, burberry root and verdigris (copper acetate). Color was achieved by applying pigments as well as by producing a chemical reaction with strongly acidic vehicles. Finishes such as spirit varnishes sometimes were colored with wood chips and bark (oak, chestnut, walnut and sumac).

Most early stains were highly fugitive. In other words, their appearance has changed almost beyond recognition since they were first applied. Thus, the main sources of information about them are cabinetmakers' and finishers' guides of the period. It is difficult to identify these stains even by sophisticated scientific techniques such as optical microscopy. The discovery of aniline dyes in the third quarter of the 19th century made coloring wood a much more precise and predictable practice.

Victorian woodwork often was stained to look like expensive hardwood. For instance, one popular treatment was to stain birch to imitate maple and other woods — mahogany and walnut were frequent choices. Victorian woodwork also was dyed by chemical processes including acid washes that turned the wood black. For Craftsman and Prairie School woodwork, "natural" processes such as fuming (coloring wood by exposing it to ammonia fumes) were preferred to applied color. The result was thought to have the appearance of age and mellowness without altering the wood's inherent character. In his book *Craftsman Houses,* Gustav Stickley recommended similar "natural" processes for coloring other types of wood. For instance, maple could be given a gray tone with a weak solution of iron rust, then sanded, coated with darkened shellac and waxed.

DEVELOPING A REHABILITATION STRATEGY

A number of issues should be considered before starting work:

- Condition and nature of the existing finish
- Appropriate treatment, both for the historic character of the interior as well as for its planned use
- Type of wood if refinishing or replacement is required

You can do an initial evaluation of the condition of the wood and existing finish yourself (see Diagnosing Wood and Finish Problems in this chapter). You can even do much of the work of reviving the finish if it is in sound condition. Where refinishing is required and a historically appropriate finish must be determined, you should call in a restoration craftsperson knowledgeable about historic finishes or a furniture conservator. If you have hired a preservation architect, he or she may consult with a specialist.

In evaluating the woodwork's condition, remember that you should look at it from the perspective of preserving as much of the historic material as possible. That means reviving the existing finish by cleaning and touching it up if it is in sound condition. Refinishing — removing the existing finish and applying a new finish — should be done only when more conservative measures prove unsuccessful or if the existing finish is determined to be inappropriate, such as a clear finish on woodwork that was painted or vice versa. If you do not know what was there originally, it is better to leave the existing finish in place. It is important not to im-

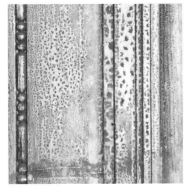

Degraded clear finish requiring stripping. (Kara Newmark)

pose your own idea of what is historic — this is interior decorating, not preservation or restoration.

Respectful rehabilitation also means patching severely damaged or rotted areas rather than removing and replacing woodwork. As in all aspects of historic preservation work, the motto is: repair rather than replace. And if replacement of a distinctive feature is necessary, match it in design, color, texture and, when possible, material (see Using Substitute Materials in this chapter for a discussion of when, where and how other materials may be used). If no physical or pictorial evidence exists for re-creating the original or later historic feature, such as molding and paneling, replacement woodwork may be a contemporary design that is compatible with the interior, or it may be an appropriate historic period design. Preservation advice should be sought from consultants familiar with the Secretary of the Interior's Standards for Rehabilitation to help address concerns regarding replacement design.

Evaluating Your Woodwork

The first question to ask is whether the wood really needs to be refinished or repaired. The best strategy in rehabilitation work is to begin with the treatment that is least destructive to the existing materials.

Finishes

If the existing finish is essentially intact — solid, smooth and not severely alligatored or degraded — it is possible that a good cleaning and some touch-up work with wood putty, a stain and a paintbrush may be all that is required. Woodwork that is grimy, cloudy, stained, scratched or coated with too many layers of wax often can be revived, sparing you the trouble and expense of refinishing.

Before hiring a refinishing contractor or arming yourself with stripper and scraper, try cleaning and touching up the existing finish (see Diagnosing Wood and Finish Problems in this chapter). You can do this yourself or hire a restoration craftsperson to do it for you. In any case, it is important to try any solution on a hidden or discreet test patch before proceeding. Depending on the finish, some of the treatments may turn it blotchy or cloudy.

Dark finishes may not require stripping either. Their color may be intentional rather than the result of age and degradation. Dark finishes were popular for various styles, particularly in the late 19th century. Finishes were sometimes pigmented to enhance or disguise woods. Undistinguished woods such as pine were often "mahoganized" in the 18th, 19th and even early 20th centuries. Stripping off a dark finish that you consider unattractive may only reveal light unattractive wood, without the character intended for your interior. Or you may find that the original stain has penetrated so deeply it is almost impossible to remove. Cleaning the woodwork, removing the top layer of finish, and checking hidden areas that retain their original color because they have not been exposed to sunlight, wear or later finishes are ways of determining if the dark color was deliberate or is a result of degradation.

Dents and damages

As for surface dents and scratches, think of them as hallmarks of age, not damage that requires repair. They offer a record of the life of a building and its occupants. Unless the wood is severely disfigured, repairs are usually unwarranted. Damage that should be patched, however, includes rotted wood, holes, breaks, deep gouges (more than ⅛ inch deep) and areas of missing wood.

Afton Villa, St. Francisville, La. The dark finish on the woodwork in this Gothic Revival interior was popular for this period. (HABS)

DETERMINING THE ORIGINAL FINISH

Determining an original or historic finish can be extremely difficult, even with sophisticated scientific techniques. The best strategy is to consult furniture conservators or craftspeople skilled in the restoration of historic woodwork. Based on their knowledge of historic finishes and evidence supplied by examination of your woodwork and relatively simple scraping and solvent tests, they should be able to advise you on appropriate finishes, stains and sheen. Most important is whether the wood originally had a clear or a painted finish. They may try sanding or dry-scraping with a knife down to the layer of finish closest to the wood. If a clear finish is discovered, solvent tests should be able to determine its composition (see Determining the Existing Finish in this chapter).

DETERMINING THE EXISTING FINISH

SHELLAC

Characteristics	Test
Made from resin derived from the Asiatic lac beetle dissolved in alcohol. One of the most popular clear finishes from the early19th century on. Classified as a "spirit varnish," it cures by evaporation of its solvent, alcohol, and can be reliquefied or dissolved with application of alcohol.	Softens or liquefies with the application of denatured alcohol.

LACQUER

A synthetic finish with a nitro-cellulose base introduced in the early 20th century. Developed for fast application by spray equipment in commercial finishing. Also classified as a "spirit varnish," it can be reliquefied or dissolved by its solvent, lacquer thinner.	Softens or liquefies with the application of lacquer thinner.

VARNISH

A solution of natural or synthetic resins in drying oils. Modern varnishes differ from their 18th- and 19th-	Probably a varnish if a finish does not dissolve with denatured alcohol or lacquer thinner.

Characteristics	Test
century counterparts. Classified as a "reactive finish": varnish cures through chemical reactions, so it cannot be reliquefied. Once degraded, varnish must be removed and replaced.	

WAX

Characteristics	Test
Usually a combination of waxes (animal, vegetable or mineral) with a solvent added for workability. Beeswax finishes were popular from the mid-18th to the early 20th century. Can be used alone or as a final protective coat on other finishes.	Softens or liquefies with the application of mineral spirits.

OIL

Characteristics	Test
Includes nondrying and drying oils such as linseed, walnut, poppyseed and tung. Popular from the 17th century on, although modern drying oils are often modified with resins or other materials for stability. Classified as reactive finishes, they cure by a series of chemical reactions and cannot be reliquefied.	Sophisticated scientific tests such as selective staining under a microscope are required.

Original color of clear finish revealed when historic hardware removed. The surrounding finish has darkened and degraded with exposure to light. (Kara Newmark)

This is a tricky procedure, however. The original finish inevitably will have changed from age, sunlight, dirt, unstable pigments or the application of later finishes. (It even may have been removed.) Also, finishes can cross-link and bond to each other so that they become difficult to isolate and identify. In some cases, a clear finish such as shellac may have been used as a sealer between the wood and an original finish coat of paint. In these situations, only microscopic analysis will reveal whether there is a layer of dirt on the clear finish, indicating that it was exposed for a period of time as a finish coat. If no dirt is present, the finish probably was painted over soon after its application and thus was used as a sealer coat.

USING SUBSTITUTE MATERIALS

While the preferred choice in rehabilitating woodwork is always to use the same materials as the original, sometimes modern substitutes may be necessary. A finish like the original may be no longer manufactured or circumstances may demand a product of improved durability. Or, when reproducing historic woodwork, a species of wood or molding profile may not be available any more.

Finishes

If refinishing is required, your first consideration should be to re-create the existing (or original, if known) finish with the same materials. If a degraded shellac finish is removed, it should be replaced with shellac. In some cases, however, considerations of durability, maintenance, renewability or environmental restrictions may suggest the use of substitute materials. The determining factors should be their effects on the historic character and on the wood itself. It is important to try to achieve the appearance of the appropriate finish and to make sure that any new finish is reversible. In any case, samples of the existing finish should be saved on the woodwork in each room for future reference.

While only shellac looks like shellac — and the same is true for lacquer, varnish and other finishes — various combinations of modern products and application methods can achieve a color, texture and luster similar to historic finishes. An experienced restoration craftsperson can be an invaluable source of advice, both for determining the original finish and for re-creating it with appropriate materials.

Restrictions currently being implemented by the Environmental Protection Agency undoubtedly will affect many of the traditionally used stripping and refinishing products. Using a system measuring the volatile organic content of each product, the EPA is encouraging the development of new water-based products that have fewer contaminants than their traditional counterparts. These products are likely to replace many of the currently available strippers, sealers and finishes, but most have not been on the market long enough to have been evaluated. A good wood refinisher should have the most up-to-date information on products that are available and their appropriate use.

Molding and paneling

Re-creating original wood molding or paneling also may be infeasible because lumber of the necessary thickness or species is no longer stocked, or the expense of custom milling is beyond your current budget.

It sometimes is possible to find molding of the correct profile and detail or have it made from a material other than wood. If the molding is to be painted, these substitutes may be acceptable in rehabilitation projects. Only wood looks like wood under a clear finish, however. The important consideration in the use of substitute materials is replicating the design, scale, detail and finish of the originals.

When infill molding is required to replace damaged or missing pieces, do not throw out the remaining intact pieces. Try to patch in the new molding. Or, if too little of the original remains, at least salvage and store the fragments. The methods of application, formation of joints and tool marks contribute to the character of molding and will not be reproduced in new molding. Future generations should have the original fragments as sources of information.

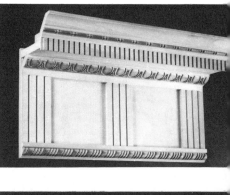

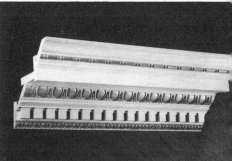

Reproduction wood cornices. (American Custom Millwork)

WHAT YOU CAN DO

If you have the time, energy and interest, you can perform most of the tasks of reviving a wood finish or, where required, stripping it and applying an appropriate new finish. If one of those ingredients is missing, however, reha-

Stair railing with poorly installed wood patch between balusters. (Michael Devonshire)

bilitating (particularly refinishing) woodwork yourself may become a tiresome chore. The cost savings and pride in accomplishing the work yourself can be more than offset by the inevitable mess, exposure to toxic chemicals and potential need to redo the work because of an unfamiliarity with products and processes.

When it comes to refinishing, professionals can do the job more easily, faster and with less disruption to your daily life. If the large scope of work or a limited budget push you toward commercial stripping and refinishing rather than hand stripping and refinishing by a craftsperson, remember that large-scale contractors may pursue the most efficient, but not necessarily most careful or appropriate, methods for your historic woodwork. If you opt for commercial methods or if you are considering doing the work yourself, it is best to consult first with a preservation architect, conservator or restoration craftsperson to determine the correct methods and materials.

Other steps in the rehabilitation process that you should delegate to professionals, unless you are particularly skilled, are wood repairs with carpentry or epoxies. These techniques require a fair amount of expertise and should not be performed by inexperienced persons, as the wood can be damaged in the trial-and-error process.

STRIPPING FINISHES

If none of the efforts to clean and revive your finish has produced satisfactory results, the wood must be stripped and refinished. This is a time-consuming and expensive procedure. It also removes the hard-earned patina of age. In other words, refinishing should be considered only as a last resort.

PAINT

Unless your woodwork had a clear finish originally, do not strip off the paint and give it a new clear finish. Trim in many old buildings from the 18th through the 20th century was softwood such as pine or fir and was intended to be painted. A natural finish for these woods is inappropriate.

DIAGNOSING WOOD AND FINISH PROBLEMS

FINISHES

Problem	Cause	Solution
Scratches	Wear, dust	Choice of solutions: rub with walnut meat; rub surface with pigmented wax; or paint with touch-up solution. Long-term: dust often; install electrostatic dust collectors.
Candle wax	Drips	Chill with ice pack; remove by hand or with plastic cooking utensil; rub surface with pigmented wax or apply touch-up solution.
White rings	Wet or hot objects	Rub with toothpaste using fine abrasive or mixture of toothpaste and baking soda or turpentine and wax.
Black rings	Moisture penetration through finish to wood	Scrape out charred wood; fill with tinted wood putty or filler sticks.
Burns	Cigarette ash, hot items	Remove top layer(s) of finish; finish may have to be stripped and stains bleached or touched up.
Stains	Spills	Choice of solutions: rub with lemon oil furniture cleanser or oil and fine pumice stone; rub with denatured alcohol; rub with lacquer thinner; or apply finish reviver.
Mildew	Condensation	Wash off mildew; expose wood to sun and fresh air. Long-term: clean woodwork regularly; install dehumidifiers
Worn areas	Wear	Touch up shellac or lacquer with new tinted finish or pigmented wax.
Bleached finish	Heat, light	Touch up shellac or lacquer with new tinted finish or pigmented wax. Long-term: install awnings and shades.

Problem	Cause	Solution
Cloudy finish	Humidity, moisture	Remove top layer of existing finish with denatured alcohol, lacquer thinner or finish reviver. Long term: install dehumidifiers.
Soft finish	Heat	Reliquefy or remove shellac or lacquer with denatured alcohol, lacquer thinner or mineral spirits.
Grime, darkness, greasiness, streakiness	Dirt, age, wax buildup	Choice of solutions: clean with soapsuds from mild detergent; remove top layer of finish with commercial cleaner or washing soda; rub with mineral spirits or turpentine; or apply finish amalgamator or reviver.
Cracked finish	Humidity fluctuations, inflexible finish	Remove or reliquefy top layer of finish with denatured alcohol, lacquer thinner or finish reviver. Long-term: install dehumidifiers.
Degraded finish	Age, light	Remove top layer of finish with denatured alcohol, lacquer thinner or finish reviver. If all layers have degraded, existing finish must be removed and wood refinished. Long-term: install awnings and shades.

Right: Water stains probably penetrating the wood finish to stain the wood itself. (Jonathan Sinaiko)

Far right: Wood paneling with degraded finish and missing moldings. (Jonathan Sinaiko)

Problem	Cause	Solution
Inappropriate quality or color of finish	Later applications of incorrect finish or stain	Remove existing finish and apply new, historically appropriate finish.
Wood character and detail obscured	Buildup of finish layers	Remove existing finish and apply new, historically appropriate finish.
Blotchy and inconsistent finish	Age, wear, light	

WOOD

Problem	Cause	Solution
Cracked or split wood	Dryness, moisture	Insert wood or epoxy patches, or fill with wood putty or filler sticks. Long-term: determine causes of damage and install humidifier or dehumidifier as appropriate.
Buckled or warped wood	Dryness, moisture	Straighten wood with wet towels and weights. Long-term: install humidifier or dehumidifier as appropriate.
Dry or wet rot	Condensation	Remove rotted area and fill with wood filler or wood or epoxy patch. Long-term: install dehumidifier or air conditioning; cover dirt cellar or crawl space with plastic tarps.
Insect infestation	Deterioration of protective finish, condensation	Remove rotted area and fill with wood filler, wood or epoxy patch; renew or reapply protective finish. Long-term: determine cause and source of infestation.
Panel shrinkage	Dryness	Glue panel in place. Long-term: install humidifiers.
Peeling veneer	Moisture, heat, dryness	Glue damaged sections. Long-term: install dehumidifier, air conditioning, shades, awnings.
Lifted or blistered veneer	Moisture, heat, dryness	Slit and insert glue. Long-term: install dehumidifier, air conditioning, shades, awnings.

Right: Painted wood finish, dining room, Amasa Day House, 1816, Moodus, Conn. (National Park Service)

Below: Alligatored paint (top, Wayne Towle, Inc.) and build-up of paint layers obscuring original detail (bottom, Michael Devonshire), both problems that warrant stripping paint.

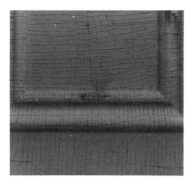

When exposed, the color and grain of softwoods are generally neither distinctive nor attractive. Also, the paint is extremely difficult to remove, usually remaining in the pores of the woodwork if the wood was not sealed first. In that case, complete removal is impossible without abrasive sanding, which will inevitably remove some of the woodwork's detail. (See the chapter Reviving Decorative Painted Finishes for how to prepare a painted surface for repainting.)

Stripping paint from woodwork is warranted, however, under the following circumstances:

- The wood originally had a clear finish and was subsequently painted.

- The paint film is unstable because of the interlayering of water- and oil-based paints or the presence of dirt trapped between paint layers.

- Detail has been obscured by many layers of paint.

CLEAR FINISHES

When refinishing, the gentlest possible stripping method should be used. Shellac and lacquer can and should be dissolved by their solvents — denatured alcohol and lacquer thinner, respectively. Varnish, however, can be removed only with a chemical paint and varnish remover. Heat guns, which are effective on paint (see the next section), do not work well on shellac or varnish. Other stripping methods also have potentially damaging side effects: scraping, sanding and some heat appliances can rip or scorch the wood.

Floor protected with tightly sealed paper and plastic tarps before stripping. (Wayne Towle, Inc.)

PREPARATIONS FOR STRIPPING

Optimally, stripping should be done outside on a horizontal surface in temperatures between 70°F and 85°F with low humidity. Professionals generally prefer to remove whatever woodwork they can without damaging it or adjacent surfaces in order to strip it in a workshop where they can control the process and work more easily on difficult-to-reach areas. Also, it keeps the mess in the shop and out of your living areas.

If the woodwork is left in place, surfaces not being stripped must be protected. The floor should be covered with newspaper, plastic sheeting and drop cloths. Walls, radiators and projecting elements such as mantels also should be covered with plastic taped firmly in place.

METHODS OF STRIPPING

A number of methods are available for stripping. As always, stripping should proceed only after a hidden test patch proves satisfactory, and a method is chosen that will not damage the wood and its details. Most experts recommend a combination of heat and chemicals for stripping paint and chemicals alone for varnish. Where there are many layers of paint, most of it can be removed with a heat plate or heat gun, followed up with chemical strippers for the residue. Heat devices are generally faster, cheaper, safer and easier to clean up after using than chemical paint removers.

Heat is less efficient at removing a base layer of paint when that layer has not been applied over a sealer coat such as varnish. Also, it is difficult to remove all the paint from details without scorching the wood, unless done by an expert hand. For these reasons, chemical strippers are generally recommended for the final layers. However, if there are fewer than four layers of paint, a chemical stripper may be all that is required.

Painted finish in the process of being stripped with chemicals. (Wayne Towle, Inc.)

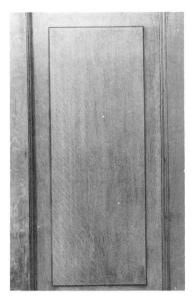 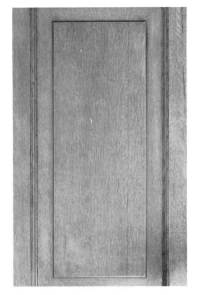 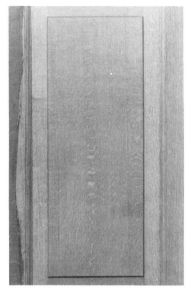

From left to right: Test panels amalgamated with commercial amalgamator leaving streaks; cleaned with trisodium phosphate and water; chemically stripped. (Jonathan Sinaiko)

Stripping paint with a heat gun. (Michael Devonshire)

Abrasives

Abrasives include hand and power sanding, spinning wire wheels attached to an electric drill and air abrasives. Sanding and wire wheels can scar and damage wood and will not remove paint from crevices. Air abrasives are useful for stripping selective layers of paint and for highly carved ornament, but it may be difficult to find a contractor or craftsperson who knows how to use them.

Heat devices

Heat appliances are not effective on shellac, varnish and some kinds of paint such as casein paint. Heat devices include propane torches, heat guns and heat plates. Propane torches can scorch wood, cause fires and volatilize lead-based paints. Heat guns and heat plates use flameless high temperatures (500°F to 750°F) and are recommended when many layers of paint must be removed. When using a heat gun, large, flat areas are usually done first, followed by carved and hard-to-reach areas. Held several inches away, the gun blows hot air onto the surface to melt the paint, which is scraped off with a broad knife or paint scraper as the paint bubbles and melts. Detailed areas require work with smaller tools such as ice picks. Professionals often use custom-ground contoured knives on long runs of molding.

Best used on broad, flat areas, a heat plate works in a

similar manner. It is held against the surface until the paint softens. Like the heat gun, the heat plate is followed by a scraper to remove the debris after the paint softens. The heat plate also can singe or burn the wood if held in one place for too long.

Chemicals

Chemical paint removers are used to strip paint or varnish. There are two basic categories:

■ Caustic removers, which are lye based and are generally used in commercial dip stripping

■ Organic solvent strippers, which are methylene chloride based and generally are preferred for interior woodwork

Strippers vary in terms of their flammability (nonflammable ones are recommended for use indoors); their consistency (semipaste or full-bodied ones work well on vertical surfaces as they will not run or drip); their rate of action (slow-acting ones are more effective where there are many paint layers or the wood is not sealed); and their rinsing liquids. Solvent-rinsing strippers are generally the most effective and least damaging to the wood, but are the most toxic. No-rinse strippers are good on overhead surfaces where rinsing liquids would drip. Water-rinsing strippers may raise the wood grain, loosen veneer and open glue joints but are less toxic than solvent-rinsing strippers.

Applying a chemical stripper.

For thick coats of finish or paint, a light coat of stripper is applied first with a brush and then a heavier coat is applied several minutes later, concentrating in corners and recesses. The surface can then be covered with plastic wrap, aluminum foil or plastic drop cloths to keep the stripper from drying out too fast. When a test scraping reveals that the paint or varnish can be removed down to the bare wood, the loosened paint is scraped up. A putty knife can be used on flat areas and a soft brass brush or stiff plastic brush in the corners. Custom-ground knives, ice picks and dental tools are used for molding and ornate areas. Carved areas and corners may need to be recoated because the paint in the grooves is usually much thicker than on flat surfaces and takes longer to soften. (Most contractors prefer to work with strippers that can be reactivated if they dry out.)

Top to bottom: Removing finish from molding with copper brush; rinsing surface to remove chemical residue with trisodium phosphate and water; lightly sanding stripped surface. (Jonathan Sinaiko)

PRECAUTIONS

Method	Danger	Precautions
Stripping and refinishing with chemicals	Exposure to toxic and carcinogenic substances	Perform work outdoors or with windows open and fans on. Wear heavy rubber gloves, goggles, a respirator, a hat and coveralls. Dispose of wastes as directed by environmental authorities. If you do the work yourself, learn the remedies for exposure and obtain manufacturers' safety data sheets from the supplier.
	Fire	Use nonflammable removers indoors. Cut off power when working near electrical outlets with steel wool.
Stripping with a heat gun or plate	Fire	Remove debris and vacuum the work area thoroughly. Seal cracks and holes that lead to studs or combustible materials. Keep a water-filled spray bottle and fire extinguisher nearby. Do not use propane torches.
	Scorched wood	Do not hold the appliance in one place for too long.
	Cracked window glass	Do not use a heat appliance on window glass, mullions or sash.
Making epoxy repairs	Fire	Keep a fire extinguisher handy.
	Toxic vapors, dust	Wear goggles, plastic gloves, a heavy plastic apron and a vapor respirator with the appropriate cartridge when applying epoxies. Wear a dust mask when sanding. Mask adjacent areas. Clean up spills immediately. Keep appropriate solvents nearby.

Wood molding, Drayton Hall, Charleston, S.C. (NTHP)

110

Wallpaper steamers are used by some contractors 10 to 30 minutes after applying the stripper. The paint's solubility is increased by the steam, which causes the paint to foam and bubble while loosening from the surface. The steam is applied through the pan of a wallpaper steamer and followed with a wide-bladed scraper. A compressed air nozzle on a steamer hose sometimes is used for detail work.

Stubborn spots may not need more attention if the woodwork will be repainted. If a clear finish is planned, however, a second application of stripper scrubbed into the wood may be required. Stains can be treated with special bleach solutions or disguised during refinishing with stain or touch-up solutions to match the adjacent wood surface.

Rinsing after stripping. Depending on the type of stripper used, close-pored woods such as cherry and maple can be rinsed with water or a solvent such as lacquer thinner, mineral spirits or turpentine. Open-pored woods such as oak, mahogany and ash usually require another application of stripper. This is then lifted from the grain with a nylon brush to remove the finish from the grain before rinsing. Once the surface is dry, the wood can be brushed with a brass brush and wiped with a clear rinse. Another strategy for removing residue from open-pored woods is to apply a mixture of shellac and alcohol to the surface. The shellac should bond to the paint in the pores, and both will be removed with another application of paint remover followed by the appropriate rinse.

Special treatments

If the paint on a test patch does not seem to budge with the application of either heat or chemicals, it may be casein, calcimine or some sort of homemade paint. Casein paint generally can be removed by scrubbing with ammonia, while calcimine paint responds to a solution of hot water and trisodium phosphate.

LARGE-SCALE STRIPPING

There are basically two choices for large-scale stripping:

- Call in a stripping contractor to do the work on site.

- Remove the woodwork and have it dipped in chemicals at a stripping shop.

Hazards of improper stripping. Scorch marks (top) from heat appliance applied too long or too closely and scrapes and gouges (bottom) resulting from use of improper tools to remove finish. (Wayne Towle, Inc.)

On-site stripping

If you choose the first option, be sure to use only a licensed and insured contractor. Stripping contractors work on an hourly, time-and-materials, flat fee or per-square-foot basis. They will provide an estimate, which is based on a test removal of finish to determine the number of finish layers, the type of paint and whether an original varnish layer is present. Before the contractor begins, you should require a test patch for methods and materials that will be used on the job. Also specify all areas to be masked.

To strip woodwork, contractors usually:

- Apply chemicals with natural-bristle brushes or airless sprayers

- Test paint regularly with a scraper or putty knife to see how much has loosened

- Scrub off the paint when the final layer is loose, then apply another layer of stripper to remove any paint residue

- Scrub ornate areas with a brass brush

- Rinse the woodwork using a high-pressure, low-volume rig

- Vacuum up waste with a wet-dry shop vacuum and store it in filter tanks for proper disposal

Stripping shops

Another method of commercial paint removal is non-hand stripping, which usually is less expensive than hiring a contractor to do it on site. This process includes flow-on, or cold-tray, as well as dip stripping. Hazards, however, include the risk of damaging the wood either with harsh chemicals or during removal and transportation. Carefully consider this option and have test pieces done at several establishments before embarking on a full-scale stripping operation.

The cold-tray method is the gentlest means of non-hand stripping. With this method, an operator

- Places a piece of woodwork in a tray and sprays on stripper, usually methylene chloride based, with a nozzle and brushes

- Removes loose finish with a putty knife and usually cleans intricate areas by hand

■ Moves the woodwork to the water-rinsing tray, where a small quantity of water is sprayed on at high pressure to remove the remaining finish and chemicals

This process is preferable to dip stripping because the pieces are not immersed in chemicals. Also, as the chemicals are reused, the cold-tray method usually is cheaper and more environmentally sound.

Dip stripping presents greater risks to woodwork than either hand or cold-tray paint stripping. It should be considered only for ordinary millwork rather than distinctive hardwoods or finely crafted woodwork. Large commercial establishments generally have three tanks, any or all of which may be used:

■ The "cold tank," which contains paint stripper, usually methylene chloride based, at room temperature

■ The "hot tank," which holds a solution of lye or trisodium phosphate and water at a temperature of between 125°F and 180°F

■ The "bleach tank," which contains oxalic acid and is used to neutralize the caustic from the hot tank and bleach out any darkening of the wood caused by the chemical stripping

The chemicals are rinsed off between dips in pressure wash booths.

Some of the problems that can be caused by dip stripping are raised grain, loosened glue joints, lifted veneer,

113

lightening or severe changes in the color of the wood, and inadequate neutralization. Most of these occur in the hot tank unless supervised carefully: the hot aqueous solution raises the grain, and the caustic solution can damage glues as well as the wood surface itself.

Although the cold-tank method is gentler, the absorption of chemicals and water rinsing can still produce raised grain and swelling. Also, the chemicals may not be thoroughly rinsed out, causing problems later. Some experts recommend having the contractor perform pH tests to ensure adequate neutralization. Otherwise, the chemicals may bleed out of the wood and cause paint or finish failure.

When trying to locate a good stripping shop, begin by asking local antiques dealers for recommendations. Call the shops and ask what treatments they use. Companies that offer the cold-tray or cold-tank method are preferable, but you should send around test pieces before you commit yourself. The stripped pieces should be checked for loose glue joints, rough wood, excessively light color (some lightening is inevitable, however) and whether finish still remains in the details and recesses.

REMOVING WOODWORK FOR STRIPPING

Removing and reinstalling woodwork may be one of the hazards of non-hand stripping, as pieces can be easily damaged or misplaced in the process. Follow these steps to protect the wood:

■ Photograph and sketch the walls from which the woodwork will be removed and key each piece in the drawing.

■ Remove hardware.

■ Label individual pieces of woodwork so that they can be easily reassembled; use a nonremovable method such as numeral dies, particularly if the pieces are to be dip stripped.

■ Repair all splinters and splits before removing the woodwork; loose pieces will be difficult to match up later, and damage may be worsened by the stripping process.

The construction at the corners of the woodwork will determine the order of removal. Inside corners of molding are usually coped, i.e., one piece was cut with a coping saw to fit the contour of the other. The coped piece is removed first, pried away from the wall with a screwdriver, hammer or prybar while protecting both the wood and the wall surfaces with putty knives or pieces of wood. Nails left in the woodwork will scratch the trim in transit, so they should all be removed or at least trimmed. When packing the woodwork, the pieces should be padded well, particularly if they are softwood. Smaller pieces should be packed in a separate box so they will not get lost.

REPAIRING DAMAGED WOOD

While the natural accumulation of marks and minor dents add character to old woodwork and thus do not call for repairs, more severe damage requires attention. Problems of this sort include holes, deep gouges, splits, breaks, rotten areas, missing pieces, loose panels, loose veneer and warpage. Preferably, these should be repaired before refinishing so that the glue residue and surface damage caused by the repair can be corrected in the refinishing process.

Professionals generally recommend that splits, breaks and loose pieces in woodwork be glued rather than nailed or screwed. Nails can split old wood, which is usually dry and brittle. They also tend not to hold over time. Where a mechanical fastener is required, wooden dowels can be used instead of nails or screws. To achieve good adhesion, the piece may have to be removed so that the old glue, dirt and paint can be cleaned off by scraping, sanding or flushing with hot water. If splits are caused by nails, the nails should be removed or set and the splits repaired.

After applying new glue, the pieces are clamped together to create enough pressure to force the glue into the pores of the wood and to keep the parts from moving. Cauls, flat sheets of wood or metal, are inserted between the clamps and the woodwork to protect the wood. If the woodwork cannot be clamped, cauls are sometimes nailed with brads over the patched area to apply even pressure. The nail holes can be filled when the cauls are removed.

Nail holes, gouges, dents and joints should be filled before the final finish is applied but usually after any necessary staining and sealing. Putty, which does not expand or contract, or pretinted wax filler sticks can be used for these

Poorly matched wood patch (below) and damage in wood paneling (bottom) requiring patching. (Jonathan Sinaiko)

115

tasks. The stick is rubbed back and forth across the hole until it is filled and smoothed. Small rotted areas can be cleaned out and filled with wood filler or colored putty. Larger holes can be filled with plastic wood. If your wood was stripped with chemicals, the original filler at joints and nail holes must be retinted or removed and replaced with new tinted filler.

Professional woodworkers may use a technique called "burning in" to fill holes, chips and scratches. Lacquer sticks are melted with a heated "burn in" knife, dripped into the hole, leveled with a felt block, sanded with a lubricant and sealed. Professionals may steam out dents with several drops of water and the hot point of a knife or with an iron over a damp towel.

WOOD PATCHES

Rotted or missing areas more than ¾ inches in diameter must be patched with wood pieces or epoxies (see Epoxies in this chapter). Carpenters use a range of wood patches, depending on the type and extent of damage:

Dutchmen are pieces of new wood inlaid into voids left by the removal of damaged wood. If the wood is to have a clear finish, dutchmen should match the adjacent wood in species, color and figure as closely as possible and be cut and fit precisely. It is best to make a dutchman from a scrap of the existing woodwork. If the wood is to be painted, however, the same species of wood should be used, but the color and grain do not have to match.

Scarf joints, usually formed by cutting long miters, join a piece of new wood to the rotted end of the original wood.

Ostrich-feather patches, used to fill cracks and checks, are long tapered slivers of wood of the same species as the wood to be repaired.

Wood patches are cut to the required size and traced on the area of wood to be repaired. A hole for the patch is then cut with a chisel, saw, drill or rasp. Glue is applied to both new and old wood, and the new piece is inserted and clamped. When set, the patch is filled or sanded to make the surface flush with the adjacent surface.

Missing or damaged veneer similarly can be patched by squaring off the patch area with a knife or razor blade, cut-

Dutchmen repairs in wood molding. (*Fine Homebuilding*)

ting a patch to fit, gluing both surfaces and clamping or weighting the patch. If the veneer is loose, which happens when it is exposed to moisture or heat, it usually can be glued down again and clamped in place until the glue has set. If the veneer has blistered or bubbled, however, the veneer must be slit so that glue can be inserted and the area is then clamped.

EPOXY PATCHES

Epoxies, consisting of liquid resins and hardeners that form a plastic material when mixed, offer an alternative to replacing damaged wood with wood patches. Epoxies are sometimes easier, faster and cheaper than carpentry if a

Wood sill before and after patching with epoxy. (Abatron, Inc.)

woodworker is skilled in their use and if enough wood remains to be consolidated, or stabilized, and patched. Large areas may be more cheaply patched with wood, however.

Unlike wood, epoxy patches are nonstructural and can be used only where support is not needed. Their advantages are that they can be easily molded to match adjacent material (some products are suitable for casting whole moldings) and they expand and contract with wood, so they will not fall out as the moisture content of the wood varies.

For woodwork of particular historical or architectural significance, epoxies preserve more of the original material and are used by museum conservators. If the wood is to receive a clear finish, however, epoxy repairs usually are more visible than wood patches and must be grained by hand.

REFINISHING STRIPPED WOOD

Refinishing includes a range of processes: sealing, staining, filling pores and finishing. Not all of these are applicable to all old woodwork, however. The nature and condition of the wood and the appropriate effect for your historic molding or paneling will help determine the best combination of finishing products and their order of use. Whether or not you are planning to do this work yourself, you would be wise to consult with a restoration craftsperson or conservator on how best to achieve a finish similar to the historic one.

Before refinishing, the woodwork should be examined for paint residue in the corners, hazy film in the open pores, raised grain, loose joints or other damage. These problems should be corrected before proceeding. They will be much harder to fix later, and clear finishes will magnify them if they are not corrected. Wood grain raised in the process of stripping can be smoothed by sanding. Sealer applied to hardwoods before sanding can stiffen the wood fibers and make sanding more efficient.

At least two to three weeks' drying time (more if the weather is damp) is generally allowed between stripping and refinishing woodwork. If the woodwork has been dip stripped, a wait of several months is recommended, while surfaces stripped of shellac by denatured alcohol can be refinished almost immediately.

SEALERS

Sealers traditionally were used to keep finishes on the wood's surface instead of soaking into the wood's pores. They allow the even penetration of stains and finishes and form a bonding layer beween the wood and the finish. If the wood grain is not set before stains or finishes are applied, dark spotty areas will occur where more stain or finish is absorbed. New wood and heavily sanded old wood should be sealed before being stained or finished. Stripped wood that was previously finished generally does not require this step. However, the need for a sealer depends on the condition of the wood, its response to stripping and what steps are to follow.

Different sealers are appropriate for different stains and finishes. Most experts recommend using a very dilute coat of the final finish as a sealer — thinned lacquer for a lacquer finish, thinned polyurethane for a polyurethane finish and so on. Diluted shellac is one of the most common sealers as it dries quickly and generally does not cause stains to bleed. (Shellac and most stains have different solvents.)

STAINS

Staining wood after stripping may be necessary for the following reasons:

- To darken wood bleached by stripping

- To make wood a historically accurate color, if it was altered by subsequent finishing

- To even out color differences, particularly where different generations or species of wood exist side-by-side

- To mask some defects such as patches, bruises, nicks and gouges

In general, wood should be stained after sanding and sealing, followed with another coat of sealer and then the finish. Most stains can be applied with a brush, a rag or even a paint roller. To prevent the stain color from bleeding into the finish coat, the stain should have a different solvent from the finish or be sealed with a diluted coat of the finish.

There are basically two ways to color wood: with pigments or by dyeing. The first process involves the application of a colored substance to the surface of the wood. The second changes the wood's intrinsic color either by diffusion of dye into the wood's cells or by a chemical reaction with the wood fiber. Pigment and dye stains can be oil, water or alcohol based.

Pigment stains, consisting of finely ground colors suspended in a vehicle such as oil or shellac, sit on the wood surface or in the pores. They are relatively color fast, but must be protected by an abrasion-resistant finish coat because they can be scratched easily. Pigments are opaque, and thus tend to mask the wood grain, although their degree of opacity varies with the pigment content. They are usually applied with a brush or rag and can be lightened by wiping off some of the stain, as they stay on the surface and, if oil based, are slow to dry.

Dye stains include chemical and aniline dyes. Such stains penetrate the wood cells and actually dye the wood. Unlike pigment stains, dyes can produce brilliant, transparent colors, enhancing the natural grain of the wood. However, they are more difficult to use than pigment stains, can become blotchy, tend to fade and can be removed only by sanding or bleaching. This irreversibility makes them problematic for use on historic woodwork. With the application of an isolating layer of sealer, however, aniline dyes in a shellac medium are frequent choices by conservators for restoration work.

Chemical stains are compounds that react with the wood to create color-fast tints without dyes or pigments. These include solutions such as ammonia, which fumes wood by reacting with tannic acid to produce a warm brown color. Other chemicals used to dye woods are potassium dichromate and ferrous sulfate. Most of these chemical reactions can be ongoing — the wood continues to change color after exposure to the chemical — and unless the wood has turned a significantly different color from what was historically intended, it is best to leave the color alone and concentrate attention on renewing the surface finish. If necessary, most of the chemical stains used historically can be reversed or repeated.

In choosing a stain for your historic woodwork, it is important to try to determine the original stain or finish color and use that as your guide. Unfortunately, many early stains are fugitive and may have changed beyond recognition because of age, exposure to light or reactions to later applications of finishes. You sometimes may find color close to the original under hardware or in places hidden from light such as undersides, backs or edges of woodwork. Also, when wood is stripped, at least some of the original stain probably will remain in the pores, providing direction on how to refinish the wood. In any case, it is best to consult with someone knowledgeable about historic stains and finishes to interpret evidence of original color and to determine appropriate treatments for your woodwork.

In restoration, sealing wood with an isolating layer of a finish such as shellac is standard practice when the wood is to be stained. The sealer allows the stain to be removed without changing the color of the wood. You should consider this step in rehabilitation as well, particularly if you have not been able to determine the original stain color, because it allows any decision you make to be reversed.

FILLERS

Depending on how it was stripped, the grain of open-pored wood (oak, ash or mahogany) may require filling with paste wood filler to create a smooth surface for an even finish. Filling pores may not be desirable, however, if the pores were not filled originally. If an open-grained appearance was intended, as in Craftsman woodwork, the pores should not be filled when refinishing.

Some manufacturers recommend filling pores after staining, others before. In either case, the process is as follows:

■ The filler is tinted to match the wood color and mixed to the consistency of cream.

■ It is brushed into the pores, and the surface is rubbed against the grain with a coarse material such as burlap or a cork block to force the filler into the pores.

■ The wood is then sanded lightly with fine sandpaper when dry.

Stripped softwood with raised grain requiring filling and sanding in order to achieve a smooth surface. (Michael Devonshire)

There are two basic types of clear interior wood finishes: penetrating and surface or film forming.

Penetrating finishes

Penetrating finishes, the most common clear finishes on 18th-century woodwork and popular through the early 20th century, soak into the pores of wood instead of forming a film on the surface, producing a sheen ranging from dull to satin. These finishes include drying oils and wax.

Drying-oil finishes, natural oils that dry to a relatively hard film, emphasize the intrinsic character of the wood, bringing out variations in color and grain. The most common are linseed, tung and walnut oils.

Linseed oil, sometimes mixed with turpentine and other oils or wax, was a common finish for 17th- and 18th-century woodwork when surface finishes such as shellac, varnish, and poppyseed and walnut oils were too costly for many people. While a linseed oil finish had the disadvantages of darkening with age and being neither water nor alcohol-proof, it was cheap, widely available and easy to apply and repair with fresh oil.

Horizontal boarding with a clear finish in a colonial interior. (NTHP)

It is difficult to identify a linseed oil finish or to distinguish it from other oil finishes, even by sophisticated tests such as infrared spectrometry and gas chromatography. If it is determined that your woodwork has been finished with linseed oil, the finish can be rehabilitated with finish reviver, but it may have to be completely stripped with paint and varnish remover if it has degraded. Boiled linseed oil, mixed with turpentine and varnish to speed drying and lessen tackiness, can be applied as a finish on woodwork. Because linseed oil finishes have so many disadvantages, some professionals recommend approximating the appearance of a linseed oil finish with a handrubbed resin varnish.

Another common oil finish is tung oil, which can be found in plain or polymerized forms. Both dry harder and are more moisture- and abrasion-resistant than linseed oil. Many tung oil products are not pure tung oil but are usually thinned or modified with resins for wood finishes. Penetrating resin oils have more predictable drying characteristics than natural oils.

Waxes can be used alone as a finish or as a final coat on other finishes not affected by its solvents. Historically, they have been used both ways. Beeswax finishes were popular, particularly on elaborately carved woodwork, from the mid-18th century to the 1930s. Waxes can be colorless or tinted and range in hardness from soft paraffins to hard carnauba waxes. Finishing waxes are generally compounds of two or more waxes dissolved in a solvent (mineral spirits or turpentine) to increase their workability. The chief waxes used in finishes include beeswax, carnauba and microcrystalline. Pure beeswax is often combined with carnauba wax to provide a penetrating, nonbrittle finish.

Waxes should only be used alone where moisture and heat resistance are not required. Otherwise, they can be applied to protect other finishes such as shellac and lacquer. Finishing waxes generally come in a paste form and are applied and rubbed with a soft cloth and then buffed. All excess wax should be buffed off, as it creates wax buildup without increasing protection.

Although they are easy to touch up or remove and can increase the water- and shock-resistance of finishes, waxes themselves will spot or soften from heat or prolonged exposure to moisture. A good choice is a paste wax containing carnauba that is neutral or tinted to your wood color. Pigmented waxes can be applied to hide scratches and mask worn areas.

Surface or film-forming finishes

Surface or film-forming finishes include lacquers, shellacs and varnishes.

Shellac is made from a resin, derived from the Asiatic lac beetle, that is dissolved in alcohol. Although shellac made from other resins such as sandarac was used on 18th-century woodwork, it was not until the early 19th century that shellac became a common finish. It remained one of the most popular wood finishes until the introduction of modern varnishes and nitrocellulose lacquer.

Shellac is still an excellent finish for historic woodwork. It can produce a range of sheens, from flat to the high gloss preferred on 18th-century and Victorian woodwork. It is easy to renew or touch up with denatured alcohol or new shellac and, when necessary, can be removed easily without damaging the wood underneath.

Terrace Hill, Des Moines, Iowa. Italianate interior woodwork with a high-gloss clear finish popular for Victorian interiors. (National Park Service)

Shellac's other virtues are its fast drying time and flexibility, allowing many coats to be applied without cracking and providing resistance to shocks caused by items dropped or knocked against it. However, it becomes brittle and darkens with age and has low resistance to water, heat, alcohol and abrasion unless waxed or varnished. If applied and cared for properly, however, shellac can have a long lifespan.

Shellac can be brushed or sprayed on. It comes in ready-to-use solutions as well as flakes or buttons to be mixed with alcohol. Modern shellac is available in four grades or colors: button lac (dark brown), orange, blonde or white, representing increasing grades of refinement. Of the four varieties, orange shellac is the most commonly used for refinishing. The "cut" of shellac refers to the ratio of lac flakes to alcohol: a 3-pound cut represents 3 pounds of lac flakes to 1 gallon of alcohol.

Lacquer, a synthetic finish with a nitrocellulose base, was developed in the 1920s and 1930s. Its fast drying time and ability to be sprayed made lacquer quickly popular as a commercially applied finish, and it remains so today. Like shellac, lacquer is easily touched up and is reversible, thus making it a good choice for historic woodwork of the early 20th century and later.

A lacquer finish also should be considered if you have a large quantity of woodwork that requires commercial stripping and refinishing. While lacquer tends to be thin, brittle and easily scratched, it can be waxed for increased protection. It is a tough and durable finish and is more heat- and water-resistant than shellac. Lacquer can be sprayed or brushed on.

Varnishes are reactive finishes composed primarily of resins and oils. Historic varnishes used resins such as copal or amber, mixed with drying oils such as boiled linseed, poppyseed or walnut oil. It was not until modern materials and manufacturing methods made possible varnishes with faster drying times that they were widely used on a commercial scale. Although more resistant than shellac and lacquer, varnishes craze when exposed to prolonged sun or heat, darken over time and can build up with too many coats, obscuring the wood's character and detail.

Modern varnishes fall into three categories:
■ Nonplastic varnishes provide a traditional, durable finish without a coated look. They produce a range of sheens from

flat to high gloss and are fairly resistant to water, liquids and abrasion. Of all the varnishes, those derived from resins provide an appearance closest to historic varnishes. Because they dry slowly and require handrubbing for the best effect, they generally are not used commercially.

Rub-on varnishes such as tung oil provide more gloss than penetrating resin oils. Like other varnishes, they demonstrate a relatively high resistance to water, alcohol and acids but are difficult to remove if they darken with age and are not usually recommended for historic woodwork.

■ Plastic varnishes are made with synthetic resins and include urethane and polyurethane. They dry quickly and produce the most durable, protective and easily maintained finishes. Their sheen can range from flat to high gloss. Depending on how plastic varnishes are applied, they can have a handrubbed look or resemble a plasticlike film that obscures the character of the wood. Although they have a relatively long lifespan, they cannot be renewed once they have degraded. As with their nonplastic counterparts, woodwork first must be stripped and refinished.

■ Water-based varnishes do not pollute the air with hydrocarbons as solvent-based varnishes do. Because they contain a much lower percentage of solids, they dry faster than their solvent-based counterparts, although it usually takes several thin coats to equal the dry thickness of one coat of regular varnish. Also, because they contain about 70 percent water, the first coat usually raises the grain of the wood, requiring that the surface be sanded. Ultimately, water-based varnishes produce a finish similar to their conventional counterparts. Their alcohol, heat and water resistance may be slightly less, however.

REINSTALLING WOODWORK

Before woodwork is reinstalled, splintered edges should be sanded; splintered or broken pieces should be glued together and clamped; and debris should be cleaned out from the joints, using stiff taping knives, dental picks or sharp chisels. The joints will not close tightly and the wood may split if crumbs remain.

The trim can be positioned and attached to the wall

temporarily with finishing nails. The nails are then hammered (preferably in existing nail holes) until they are one or two taps from the surface. A nailset is then used to set them slightly below the surface, as hammer taps will damage the wood. The nail holes can be filled with wood putty and sanded when dry.

RE-CREATING MISSING WOODWORK

I f pieces are missing from existing woodwork or if all the original trim has been removed, you may want to consider installing replacement molding or paneling if you have documentation for the original design. First search the premises for original remnants — check the attic, closets, garage, carriage house, barn and other places for pieces that match. Another possibility is to trade matching woodwork in better condition in nonpublic rooms with damaged woodwork in public rooms. You may also be able to find pieces with the same profiles at a local salvage warehouse. Failing these possibilities, however, you may have to resort to new millwork. Epoxies also can be used to reproduce infill pieces or entire moldings.

Door trim with missing corner block that must be replaced or re-created. (Jonathan Sinaiko)

RESEARCH

If there are no traces of the original trim or original drawings, but there is evidence that molding or other woodwork did exist in your interior, how do you choose appropriate replacement molding? The best advice is to consult a preservation architect, architectural historian or restoration craftsperson familiar with the history of local woodwork and molding types and profiles.

They may begin by identifying a group of buildings in your area of the same type and date. Examination of trim in those buildings and identification of their builders and artisans may suggest common features for appropriate designs. Particular molding profiles and proportions may have been popular at certain times, and documentation of these may provide information for your own woodwork.

Architectural pattern books of the period also are useful but should be reviewed by an architectural historian or preservation architect knowledgeable about local tastes and customs. These books provide a wide array of choices, and profiles shown are often too formal or elaborate for the typical buildings that most people own. The up-to-date styles and methods shown in pattern books also may not have been followed in your particular area or by your particular builder. In the absence of documentation, it may be best to install a simple contemporary molding — or none at all.

MOLDING

Once you have determined the appropriate molding profile or if a portion of the original trim remains, you may be able to find new stock molding to match. Many patterns have a long shelf life, so the profile may still be available, if not locally, then from specialty suppliers. It also may be possible to assemble the desired profile from several stock moldings. Wood of the same species and approximate age as any existing woodwork should be obtained if the pieces are to have a clear finish.

If neither option yields success, molding can be custom milled based on an existing piece or drawings of it. (Some companies accept mail orders if this service is not available in your area.) Molding machines use one or more knives or cutters. Because molding is usually a combination of several sections, a combination of knives probably will be required. An old mill may have the appropriate ones in stock but will charge a set-up fee. If the knives are not available, however, they must be ground to order. Carpenters also can make molding and have knives custom-made for their use. Or molding can be made the traditional way with hand planes, which can be ordered from specialty suppliers or purchased from flea markets or antiques stores.

Recording molding profiles

Before acquiring or commissioning new molding, accurate data about the existing molding should be collected. Exact dimensions, the species of wood and the profile should be verified on a clean, undistorted, undamaged piece from which all the layers of paint have been stripped. It is

Original moldings revealed behind later wood paneling. Missing or severely damaged pieces will need to be replaced with moldings matching these. (Jonathan Sinaiko)

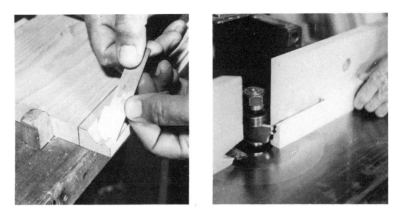

Re-creating historic molding. Left to right: Recording molding profile with profile gauge; custom-ground knife, or cutter, to fit existing profile; vertical spindle power shaper with cutter before installation of safety guards, guides and power feeders. (John Leeke)

best to remove a sample of molding from the wall to take to the millwork shop, as the thickness of the wood and the moisture content are probably different from today's standard milled wood.

If the existing molding cannot be removed without damage, the profile can be recorded with a profile gauge — a series of fine pins, bars or plates that slide back and forth to fit the contours of a piece of molding. To record a profile, the gauge is pressed against the molding at a right angle; the displaced bars create a negative impression of the profile, which can then be traced on paper.

Profile gauges cannot record severe undercuts, however. Other tools sometimes used are flexible wires or sheets that can be pressed into the profile and traced. These include soft-cored solder, thin lead sheet, draftsmen's flexible curves or many layers of aluminum foil. However, most of these tools have a tendency to lose their shape when removed from the model and can miss sharp changes of angle. For complex moldings with heavy undercuts, molds of rubber latex or silicone rubber reinforced with gauge bandage can be made.

Installing replacement moldings

Re-created moldings can be installed using standard finishing nails. Moldings at outside corners will require mitering with a miter box and back saw so they will form a right angle when joined. Moldings at inside corners must be coped. This also can be done with a miter box, with the first piece trimmed at a 45-degree angle, its cut profile serving as a guide for a coping saw on the second piece, also cut at a 45-degree angle. The two faces should then fit together. Nails are left slightly above the surface and then set slightly below the surface. The nail holes are then filled with colored putty.

WOOD PANELING

Re-creating wood paneling is more problematic than molding because many of the woods once commonly used, such as chestnut and elm, are difficult to locate. Wide boards, sometimes more than two feet in width for historic paneling, also are no longer available and must be made by gluing together narrower boards. While it is best to use the same species as the original wood, where wood is to be painted, some professional cabinetmakers recommend using birch or walnut veneer plywood, which will not expand, shrink, warp or crack as solid boards do.

As with molding, the design of new paneling should be based to the greatest extent possible on documentation of the original design. If none remains, but there is evidence that paneling once existed in a room, hire a preservation architect or cabinetmaker experienced in the restoration and reproduction of woodwork to design paneling that is suitable for the age, style and class of your building.

Each wall must be measured precisely, with windows, doors and electrical outlets and switches located on drawings. The paneling design should maintain authentic proportions while keeping the spacing of the panels consistent. The drawings for the paneling can then be given to a cabinetmaker or millwork shop as a guide to making the assembly of frames (stiles and rails) and panels.

To install paneling on a wood-frame wall, plaster and lath are removed down to the studs and horizontal furring strips installed. The paneling is then nailed directly to the studs. Moldings such as chair rails and baseboards are applied to the face of the paneling. If the wall is brick, the paneling can be assembled directly on the plaster using cut or masonry nails. If necessary, wooden plugs can be installed in the brick and wood screws used.

Severely damaged wall requiring new plasterwork and re-creation of wood wainscoting where drywall has been installed. (Jonathan Sinaiko)

MAINTENANCE

The objectives of woodwork maintenance are preservation, protection and prevention. How durable finishes are depends on environmental factors as well as daily wear and tear. The chemical nature of the finish itself, in addition to its method of application, also affects its durability. Some finishes, such as shellac and lacquer, last for decades if applied and maintained properly. Other finishes, such as drying oils and varnishes, are shorter lived because of ongoing chemical changes.

CAUSES OF DAMAGE

In addition to light, heat and moisture, even such seemingly innocent forces as dust and plants can be sources of trouble for woodwork. The causes of potential damage are many:

Sunlight and strong fluorescent or incandescent light are major sources of degraded finishes.

Heat softens finishes and causes localized problems such as bleaching, lifting veneer and tackiness.

Dryness makes wood crack or buckle and shrinks panels.

High humidity, if prolonged, causes finishes to absorb water and become cloudy, softens glue, encourages mold and makes veneer or inlay peel.

Condensation, a problem particularly on window and door frames, window sash and sills, produces paint bubbles, lifted veneer, dry rot, wet rot, warpage, blackened wood (when moisture penetrates below the finish) and a haven for termites.

Fluctuations in humidity create stresses as wood shrinks and expands.

A finish that is not elastic enough may crack because of different coefficients of expansion for the wood and the finish (see Diagnosing Finish Problems).

Ordinary dust is abrasive and scratches woodwork.

Surface dirt — dust, fibers, grease, soot and fungi — obscures the wood's character and attracts moisture that will increase oxidation and mold.

Plant pots leak, dent, scratch, leave white or black rings, lift veneer and cause splits or warping. Plants do not belong on old wood surfaces, so remove them or at least use waterproof, scratchproof pots and set them on protective pads.

You can take a variety of measures to minimize damage to woodwork:

- Dust wood often.

- Install air conditioning and electric dust collectors if feasible.

- Cover the ground with plastic tarps to prevent ground moisture from rising up through the building, if moisture is a problem and your house has a dirt cellar or crawl space.

- Wash off mildew when it appears, exposing the woodwork to as much sun and fresh air as possible.

- Combat dryness with humidifiers, which are particularly useful in dry parts of the country and in buildings heated in winter.

- Reduce damage from sunlight with awnings and shades.

- Mitigate the effects of atmospheric pollution with regular cleaning as well as air conditioning

- Apply paste waxes, the best protective coatings for finishes. They are compatible with most finishes, are relatively inert chemically, provide a good moisture barrier and can be removed easily. Commonly used alternatives such as linseed oil do not provide a good moisture barrier and can bond chemically to the finish, becoming almost impossible to remove. Lemon oil is a nondrying oil that will saturate the finish color but provides no protection and will attract dust. Liquid spray and wipe-on polishes leave a soft film that will collect dust and may contain chemicals that can react with the finish and cause softening.

- Wash most woodwork, depending on its finish, several times a year with a minimal amount of water and a mild detergent and dry immediately. Test first to make sure that the finish does not respond adversely.

- Clean wax finishes by dusting and periodic washing with a dilute solution of water-based detergent and a soft cloth.

- Rewax finishes, depending on wear, once a year, making sure that they are clean first. Dirt trapped between layers causes many of the problems associated with wax buildup: obscured detail and character, griminess, darkness and streaking.

PRESERVING PLASTERWORK

Problems with flat plaster walls or ceilings often look worse than they actually are. A few cracks and holes or missing pieces do not mean that the plaster's life is over and that the wall or ceiling must be replaced. Alligatored paint, loose canvas, soft spots, cracking, holes, delamination of the finish coat and loose or bowed plaster usually can be corrected. Damaged, broken moldings or missing pieces of plaster ornament can be restored or, where necessary, replaced.

Bulging or cracking in a plaster wall may be the result of a number of factors, including structural movement, extreme temperature or humidity changes, leaks, broken plaster keys, faulty plaster mix, improper curing or loose or rotted lath. Other problems may occur just in the paint or wallpaper layers or in bad patches applied over the years. Except for defects inherent in the plaster mix or its original application, all of these problems probably can be repaired without resorting to full-scale plaster removal.

FLAT PLASTER TECHNIQUES

Plastering is divided into two types of work: flat and ornamental. Flat plaster on wall and ceiling surfaces was traditionally applied in three coats, and the three-coat system is still considered the best method of producing the strongest and straightest plaster finish. Two coats, however, are possible, depending on the type of lath and building code requirements.

Standards for plaster thickness and straightness are set by grounds and screeds. Screeds are narrow strips of plaster, which act as guides for the straight edge, or rod, used to

Wood lath with plaster removed. (Jack Boucher, HABS)

Opposite: Creating a mold for plaster ornament for the World's Columbian Exposition of 1893. (Daniel and Smock, *A Talent for Detail*)

133

Right: Applying three-coat plaster. The depth of plaster coats is kept smooth and consistent through the use of grounds (a) and screeds (b, d, e, f), which act as guides. A two-handled float, or darby (g), is used to even out the base coat. A straight-edge (c) levels the grounds and screeds so they are the same depth.

Below: Wire mesh reinforcement, or cornerite, applied at corners to protect against cracking. (both, *Old House Restoration*)

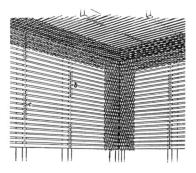

smooth the plaster surface. Set in plumbed, level lines equal to the depth of the next coat, screeds establish standards for flat, smooth surfaces of consistent thickness. When screeds are dry, plaster is filled in between them. Traditional practice also includes the installation of corner beads and grounds (wood or metal strips) around doors and windows and at the tops and bases of walls. These strips also guide the depth of plaster coats as well as serve as nailing strips for molding.

LATH

Plaster is attached to a wall by an adhesive or mechanical bond formed between the first coat of plaster and the lath. Made of a variety of materials, lath usually has many openings through which the plaster is pushed to form keys behind the lath, bonding the plaster to it. An adhesive bond also may be formed by the suction created when wet plaster is applied to lath with a continuous surface such as gypsum board lath.

Traditional lath was wood, but now expanded metal, wire, gypsum or rock lath are more common because they are easy to install. In addition, wood is vulnerable to moisture and insect damage and tends to expand and contract with humidity changes, causing plaster cracking. (Gypsum and rock lath, however, also are vulnerable to moisture damage.)

Another kind of lath, "paint-on lath," is actually a bonding agent that provides strong adhesion for plaster on a sound, clean surface. It is particularly useful for repair or patch work.

Lath types include the following:

Wood lath, narrow strips of wood nailed to the structural frame or furring strips, leaving spaces in between for plaster keys.

Metal lath, various configurations of metal wires, ribs or perforated sheets nailed, stapled or attached with wires to the framing. Because it provides more spaces for plaster keys than wood lath and is resistant to rot and insect infestation, metal lath is preferred by most contemporary plasterers.

Four types of lath. Clockwise, from top left: Hand-riven wood lath; metal lath; perforated gypsum board lath; and machine-sawn wood lath. Sections show typical plaster keys for each type. Metal lath is typically used for new or replacement plasterwork because it is resistant to rot and insects and allows for numerous keys. (Kaye Ellen Simonson)

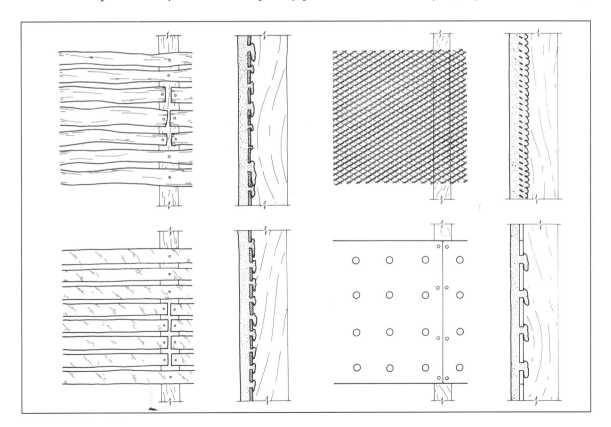

Right: Hand-riven wood lath. (Michael Devonshire)

Far right: Machine-sawn wood lath. (Jonathan Sinaiko)

Right: Metal lath. (Michael Devonshire)

Far right: Unusual straw lath. (Michael Devonshire)

Board lath, made generally of mineral and vegetable products compressed between sheets of specially prepared paper to form solid or perforated boards.

THE THREE-COAT SYSTEM

The three traditional coats of plaster are the scratch, or first or base, coat; the brown, or second, coat; and the white, or finish, coat.

Scratch coat

The scratch coat is necessary to provide a strong and stable base for the coats that follow. It forms a bond with the lath and stiffens it so it will not buckle when subsequent layers of plaster are added. Plaster is mixed in a box with a mortar hoe or by a mechanical plaster mixer, applied to the lath using a hawk and trowel, and pushed behind the lath to form keys. Fibered plaster often is used

136

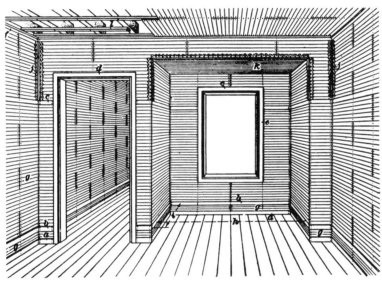

for new base coats, as the fibers help hold the plaster together and prevent too much plaster from slipping through the spaces between the lath. The scratch coat usually is approximately ⅜ inch thick. As it begins to set, the plaster is crosshatched to provide a mechanical bond for the second coat.

Brown coat

After the scratch coat has set, the brown coat is applied to build up and even out the surface. It is usually ½ to ¾ inch thick, straightened with a rod and darby, and left rough to create a good bond with the finish coat.

Above left: Applying the base coat. Plaster is held on the hawk (b) and applied with a trowel (c). This coat is then scratched with a comb (d) to provide good keying with the second, or brown, coat.

Above right: Applying the finish coat. Hawk and trowel are also used on the finish coat (a). A plasterer smooths the surface with water applied with a brush (c) and a steel trowel (b).

Left: Early 20th-century interior before plastering. Plaster grounds (wood strips) are installed at locations a, b, c, d, e and f. Metal corner beads, at locations j and k, protect outside corners and regulate the thickness of the plaster. (all, *Old House Restoration*)

137

Finish coat

Two plasterers usually are required for the finish coat — the first to apply the plaster, followed by the second to straighten the coat with a tool called a featheredge. The finish coat sometimes is applied in two coats; one fills out the roughness in the brown coat, and the other, applied while the first is still wet, is straightened with a wooden, plastic or aluminum float to create a smooth, even surface. The finish coat is allowed to set for several minutes before it is wetted again with water and troweled to polish the surface. This procedure may be repeated several times.

TOOLS

Tools used to apply plaster by hand are similar to those used in the 18th and 19th centuries. They include a hawk to hold the plaster while it is troweled onto the surface; various types of trowels to apply, spread and smooth the plaster; and floats to straighten the surface, level off bumps, fill in hollows and compact the plaster into a smooth dense finish. Textured floats are used to create different surface textures. Darbies, slickers, rods and featheredges smooth the plaster surface. Contemporary plasterers sometimes use machines to mix and apply the plaster, although the preparation, such as the creation of screeds, and finishing are the same as when plastering by hand.

Tools used for traditional plastering: screens (a, b); hawk (e); assorted trowels (f, g, h, i); float (j); darby (k), or two-handled float, used to float larger surfaces; square (m), used to measure trueness of angles; jointing and mitering tools (o, p, q), used to shape angles in moldings; comb (s) of sharpened pieces of lath used to scratch base coat; brush (t), used to dampen plaster while smoothing. (*Old House Restoration*)

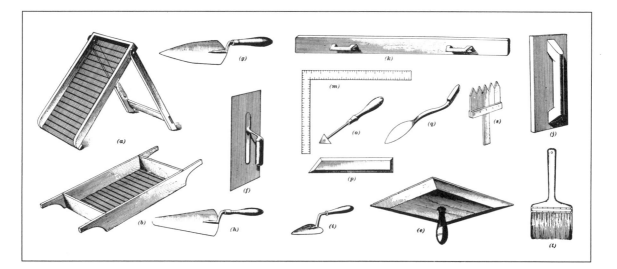

138

PLASTER INGREDIENTS

Plaster consists of a cementing material (lime, gypsum, portland cement), an aggregate (sand) and water. Sometimes binders such as fiber or animal hair are added for reinforcement. Other ingredients are used to retard or hasten the setting time.

Several kinds of plaster have traditionally been mixed for walls. The scratch and brown coats, which provide the foundation for the finish coat, traditionally consisted of about two parts water to one part lime paste, one part animal hair and two parts sand. Modern base coats generally use gypsum instead of lime in a proportion of two parts sand to one part gypsum. Fibered gypsum may be used over wood or metal lath to prevent an excess of plaster from being pushed through the openings. The third, or finish, coat contains a much higher proportion of lime: about three parts lime paste to one part gauging plaster, a specially processed gypsum. Marble dust or fine white sand sometimes was added to the finish coat to achieve particular effects.

Cementing materials

Lime putty, used for most interior plasterwork until the turn of the century, is made from lime, or calcium carbonate, derived from quarried limestone or shell deposits. It is heated to high temperatures (1,650°F to 2,500°F) to form quicklime, which is then slaked, or hydrated (mixed with water), for use in plastering. Because lime is caustic, goggles, a dust mask or respirator, and latex gloves are required to protect workers mixing the plaster. Several types of lime are currently available, including the most common ones used for interior plastering:

■ Hydrated, or regular finish, lime is sold as a dry powder and must be soaked 12 to 24 hours to make the lime putty used in plastering.

■ Autoclave, or double-hydrated, lime requires no presoaking and can be mixed with water and used immediately. Autoclave lime generally is preferred by plasterers, as regular finish lime sometimes has lumps of unhydrated magnesium that can absorb moisture from the air and eventually expand into bulges, popping the finish coat off the base coat.

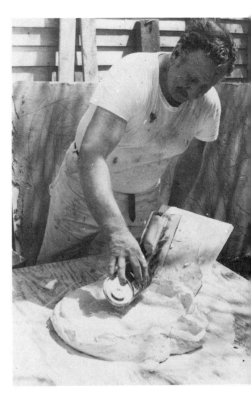

Mixing finish coat of plaster. Ring of lime putty is filled with water, which is then sprinkled with gypsum plaster. The lime putty, gypsum plaster and water are then mixed together. Gypsum plaster increases the setting time of plaster, so the mixture must be used immediately. (Mangione Plaster, Saugerties, N.Y.; John Leeke)

Plasterers mixing large quantity of lime putty finish coat. (David Flaharty)

Lime plaster can take up to a year to set, however. Its slow hardening, stringent storage requirements and tendency to shrink if not properly proportioned were responsible for its loss of favor to gypsum around the beginning of the 20th century. Gypsum was preferred because it dries quickly, makes more rigid plaster and generally does not require a fibrous binder. However, lime putty continues to be used in finish plaster, where it is combined with gauging plaster. The lime putty increases the workability and drying time of gypsum plaster.

Gypsum, currently used for most plasterwork, is, like lime, a mineral, calcium sulfate. It is refined into plaster of paris after it has been ground and heated, or calcined, to remove water. When remixed with water, calcined gypsum sets into a rocklike condition and has valuable fire-resistant properties. Although lime and gypsum plaster resemble each other visually, gypsum plaster generally does not require a fibrous binder unless applied over wood or metal lath. It also is more vulnerable to moisture damage. If applied to a masonry wall, furring strips are required to separate the plaster from the masonry to protect it from moisture penetrating the wall.

There are many types of gypsum plaster. Plaster used for walls and ceiling surfaces generally falls into two categories:

■ Unfibered gypsum, or neat plaster, also known as compound or cement plaster, has no added aggregate or filler. It must be mixed with water and an aggregate before using.

■ Fibered gypsum has a binder such as cattle hair or sisal fiber added for reinforcement. Wood-fibered gypsum uses wood fibers and is much stronger, harder and more fire resistant. It can be applied to all standard lath and masonry surfaces and is often recommended as a scratch coat for wood and metal lath.

Portland cement, although most often used on exteriors, sometimes is mixed with lime and sand for use in interior areas of hard wear or dampness. It is made by grinding, calcining and pulverizing mixtures of lime with clay and adding gypsum to control the setting rate.

Aggregates

Aggregates are added to plaster mixtures to help prevent shrinking, which would cause cracks, and to make the plaster more economical by adding filler. The traditional aggregates were sand and sometimes crushed oyster shells. More commonly used today are lightweight aggregates such as vermiculite, perlite and pumice derived from natural rock deposits. These can improve the insulation, sound-absorption and fire-resistant qualities of plaster. They are also lighter in weight than sand.

Sand, however, should be used as the aggregate when recreating historic plaster finishes. Sand must be clean, so as not to cause streaking and unevenness, and of a certain size and coarseness. It is graded according to size and moisture content, which can help determine how much water is added when plaster is mixed.

Water

Water makes the plaster mixture workable and dissolves the cementing material so that it acts as an adhesive to bond with the aggregate. The amount of water required depends on the base and plaster mixtures. For plasterwork, it should be clean and pure, as any impurities can cause discoloration or alter the setting rate.

Admixtures

Any ingredient added to the basic plaster mix (cementing material, aggregate and water) is called an admixture or admix. Admixtures are used to modify particular characteristics of the plaster mixture, such as the setting time, strength or color.

Binders are used to strengthen plaster. Animal hair or vegetable fiber were traditionally used in lime plaster to add strength and cohesiveness while it set. Fiber or hair is also sometimes added to prevent too much plaster from being pushed through the spaces of wood and metal lath. Animal hair was the common binder used until the early 20th century. Now animal, vegetable, mineral and wood fibers are all available.

Accelerators speed the setting time of gypsum and lime plasters. Gypsum itself was a traditional accelerator to quicken the setting time of lime putty finish coats and is still used today for this purpose. Other accelerators work with different cements, for example, sulphate of potash for Keenes cement. An accelerator may be required when a plaster mix is particularly slow setting or when compensa-

Top: Plaster applied directly to masonry that has been damaged by water seeping through the masonry. Damp plaster must be removed and replaced — preferably on furring strips to separate it from the masonry. (Jonathan Sinaiko)

Above: Settlement cracks that are easily repaired. Large corner crack may represent a structural problem and should be investigated further. (National Park Service)

tion for weather conditions or a particular aggregate is needed.

Retarders slow the setting time of plaster mixtures. They are often added to the finish coat so that it can be troweled smooth before it hardens. Retarders work by slowing down the absorption of moisture, which is the process by which gypsum hardens. Other retarders include cream of tartar, gelatin, glue, ammonia, zinc sulphate, soap, starch, and animal or vegetable oil.

Color admixtures were popular for use with textured or imitation plaster finishes in the early part of this century (see Textured Finishes).

DEVELOPING A REHABILITATION STRATEGY

In general, patching an existing wall or ceiling is less expensive than demolishing and replacing it. Only if more than 50 percent of the plaster has deteriorated should you consider removing it. Inspect the wall first to determine whether the problem is in the surface finish, such as the paint or canvas, the plaster itself or the substrate, such as the lath or structural frame. Loose canvas (used to hide plaster flaws) or cracked and peeling paint can be repaired easily or removed and replaced without damaging the plaster surface underneath. Similarly, random cracks and holes can be patched while preserving the plaster surface.

In some cases, however, severe damage apparent in the plaster finish may be a sign of more fundamental problems, such as ongoing leaks or unusual structural stresses, which must be addressed before any patching and repairing of the wall surface. Otherwise, the damage will probably recur. Where the damage consists of loose, bowed, soft or delaminated plaster, or large cracks where the plaster surfaces have shifted planes, you should call in a preservation architect to track down the source of the trouble.

It is also important to complete all structural, mechanical, electrical and plumbing work before you embark on any plaster repairs. This work, while done in or behind walls and ceilings and carefully planned to cause minimal disruption of historic fabric, often requires some cutting and removal of plaster for access (see Building Systems in the chapter Planning for Rehabilitation). Wait until all holes and cracks can be patched at the same time to avoid replastering sections already completed. Similarly, major structural work may cause the building to shift and settle, so allow several months for the building to stabilize before repairing plaster finishes.

EVALUATING YOUR PLASTER

An initial evaluation of the plaster's condition can be conducted by the homeowner, along with some of the simpler repairs such as removing the deteriorated painted finish and patching minor holes and cracks. Most plaster repairs, however, are best done by a professional plasterer. As you conduct your evaluation, be aware of the following conditions:

Soft, powdery plaster and cracking and loosening of the finish coat may be the result of incorrect proportions of ingredients in the original plaster mix, poorly mixed plaster or incompatible base and finish coats. Map cracking, or webs of hairline cracks, and surface delamination also may stem from extremes of temperatures or humidity while the plaster cured. For instance, freezing temperatures can make the finish coat swell and scale off: as the water content expands, it causes the plaster to burst. High humidity can cause plaster to sweat or rot from drying too slowly, while too little humidity may make plaster dry too quickly and result in the loosening of the finish coat. Similarly, the direct heat of a furnace or steam radiator may produce map cracking and separation of the top and bottom coats.

Plaster strength also is dependent on the mix and the care of its application. If applied too thinly, it may not be able to withstand movement and will crack easily. If too thick, more than ½ inch in a single coat, for instance, it may slip off the wall before setting. Also, if the finish coat was applied over an excessively dry base coat, a suction effect may have occurred, causing the finish coat to shrink and crack. Or the finish coat may crack if it is incompatible with the base coat, sometimes due to different shrinkage rates. Another problem could be created by insufficient troweling. Because troweling compacts plaster into a hard, dense layer, inadequate troweling may have left a finish coat spongy.

Cracks can result from movement within the building frame or within a particular wall. Stress cracks generally are

Top: Wallpaper and plaster damaged from leaking plumbing. Pipes must be repaired and damaged plaster and wall covering removed before holes can be patched. (National Park Service)

Above: Hole in plaster resulting from water damage. Remaining plaster is loose from the lath and will need to be reattached. (John Leeke)

diagonal, radiating from the corners of door and window frames, but they also can occur randomly within a wall. These cracks usually are caused by the structural movement or flexing of the door or wall framing when it is inadequate for the load or by changes in the moisture content. Horizontal cracks in a wall may stem from expansion and contraction of wood lath because of humidity changes. They may also occur at seams in board lath panels.

Diagonal cracks in opposite directions suggest other sources of movement, such as settlement of the building, particularly if it was constructed on unstable soil. They also can be caused by a nearby source of vibrations, such as a highway or railroad.

Water damage shows itself in brown rings, crumbling or efflorescence, a chalky powder formed when salts in plaster are brought to the surface by water. Dampness may be caused by a number of problems: leaks from gutters, downspouts or the roof itself; interior plumbing leaks; or, if the plaster was applied directly to masonry, water wicking up the walls from runoff against the foundation or dampness in the ground. If the leak is stopped quickly, the plaster may not be seriously damaged. If allowed to continue, however, the affected plaster is probably irreparable.

You can test for the extent of damage by cutting through the plaster with the edge of a putty knife. If it slices easily through the plaster beyond the finish coat, moisture has damaged the base coats. Soft spots can also be found by tapping on the wall for hollow sounds.

Loose or bowed plaster occurs with delamination of the finish coat from the base coat. It may also be caused by broken keys, as well as by lath that has come loose from the framing.

You can test to determine the nature of the problem by pressing the bowed area of plaster. If it seems to move in relation to the lath and framing, the keys are probably broken. If it seems to move in relation to the studs or joists, the lath is loose from the framing. If the latter is the case, you may discover problems of decay or structural damage as well, and a test area of plaster should be removed for further investigation.

Overall surface deterioration, or a cracked, alligatored or peeling paint surface, is one of the most common problems. This may result from too many paint layers. Distem-

DIAGNOSING PLASTER PROBLEMS

Damage	Cause	Solution
Dry, powdery	Incorrect mix, application or setting plaster time; excessively hot or cold temperatures, dryness or humidity while curing; proximity to furnace or steam radiator	Remove unsound plaster and re-plaster
Cracks	Incorrect application; incompatible base and finish coats; building settlement; vibrations; structural movement; expansion and contraction of wood frame or lath	Remove unsound plaster; patch cracks and holes
Brown rings, crumbling, efflorescence	Water damage from leaks or water penetrating masonry walls	Seal and repaint stains; remove unsound plaster and replaster.
Loose or bowed plaster	Delaminated finish coat; broken keys; loose lath; rotted lath; structural failure	Choice of solutions: apply new plaster keys; repair or replace lath; install plaster washers or inject adhesive bonding.
Cracked, peeling paint surface	Too many paint layers; calcimine paint overpainted with latex or oil-based paints; failed texture paint; loose canvas	Remove unsound paint and repaint; remove loose paint and apply canvas.
Damaged or missing molding	Removal; inappropriate alterations	Run new molding in place or on a bench or cast new molding to match the existing one.

Crumbling plaster (right) and peeling canvas (far right) resulting from water infiltration. (Michael Devonshire)

per paints such as calcimine also will cause peeling if painted over with latex or oil-based paints. Other problems may be caused by a poorly applied texture finish that is beginning to fail or by loose or bubbled canvas that has lost its adhesive.

USING SUBSTITUTE MATERIALS

As in all rehabilitation work, historic materials should be patched and saved wherever feasible. If replacement is necessary, use materials that match the historic appearance as closely as possible. For instance, a veneer plaster system looks more like historic plaster than drywall. Even if removal of an entire plaster surface is required, your first consideration should be to replace it with new three-coat plaster on existing or reinforced lath. Not only are plaster walls stronger and more soundproof than drywall or veneer plaster, their handmade texture is hard to duplicate with modern substitutes. Three-coat plaster also has valuable fire-resistant and insulative qualities. Most important, replacing a lath and plaster wall with a modern system such as drywall diminishes the historic character of an interior. Similarly, removal of a plaster finish to expose a brick wall is soundly discouraged, as it is almost always historically inappropriate.

Included in this chapter is some discussion of alternatives to plaster for walls, ceilings and ornament. Ideally, you should replace plaster with plaster. However, in some situations, particularly where plaster ornament is extensively damaged and must be replaced, the ease and relative cost of modern substitute materials may make the difference between re-creating the appearance of a historic interior or compromising with an unornamented interior. It is essential, however, that new ornament such as moldings or ceiling medallions be based on documentation of the existing decoration. If no documentation is available, but you have clear evidence that some type of ornament formerly existed, consult with a preservation architect or architectural historian to choose a replacement that is appropriate to the age and character of your building. An architect can also help you evaluate technical problems and compliance with local building codes for any new work.

REPAIRING DAMAGED PLASTER

If the damage to your plaster walls or ceiling is confined to the paint surface, there are several solutions. If the paint is alligatored, the unsound layers may be removed and the wall or ceiling repainted. Or the loose and flaking paint may be scraped off and canvas applied to the wall to create a smooth surface for repainting.

REMOVING PAINT

If you decide to remove all the paint, make sure to save samples of all paint layers on each surface in a protected, discreet place. They serve as cumulative evidence of the history of your building and should be retained for the information of future generations.

Paint usually can be melted with an electric heat plate and scraped off with a wooden or metal scraper. (Check first to make sure that a layer of wallpaper is not under the paint, because this could present a fire hazard.) Sanding or a chemical stripper will remove the residue. Chemical stripper can also be used on ornament and moldings if they are not too fragile. Paint removal techniques are similar to those used for woodwork (see the chapter Repairing Woodwork for materials and methods as well as precautions).

If an earlier coat of calcimine paint is the cause of paint surface failure, a wallpaper steamer can be used at the edges of a loose area, allowing the steam to get under the calcimine. All the layers should then be removed with scrapers. If exposed, calcimine paint can be removed with trisodium phosphate, hot water and a sponge.

Unwanted textured paint also will usually come off with the application of steam from a wallpaper steamer to break the paint's bond with the plaster. When using a steamer, the pan is held against the wall to thoroughly wet the finish, and the loosened paint is removed with a scraper. After the wall is rinsed, it may require patching and some resurfacing.

If, however, the textured surface is the result of a tooled plaster finish coat rather than paint, it probably will have to be removed and a new finish coat applied. Some textured finish plasters were applied to the existing finish coat and may be weakly bonded. These plasters can often be removed by steaming and scraping. You should be aware, though, that textured finishes were sometimes applied to cover failing plaster, and the rest of the plaster may come off with it.

INSTALLING, REPAIRING AND REMOVING CANVAS

If the plaster surface is sound but cracked or uneven, canvas (a fabriclike material similar to wallpaper) may be applied to the wall. A traditional treatment often used in historic buildings to camouflage plaster flaws, this repair method is helpful in smoothing out a slightly rough surface. Cyclical or hairline cracks also can be bridged by canvas without first being patched. Large cracks, dents, holes and severely uneven surfaces, however, should be stabilized and patched before canvas is applied. Also, areas where the plaster is actually damp or crumbling must be removed and patched (see Patching Holes). (When planning a decorative painted finish, canvas will provide a stable substrate that prevents damage from small plaster cracks. It also contributes a clean surface for paint if a wall has been damaged by fire or smoke.)

Fabrics smooth enough to resemble a plaster finish when painted are available. Some are cotton, although most currently manufactured have a synthetic base such as polyester or fiberglass. Fabric can be used in conjunction with lining paper, which will prevent shrinking and provide good tooth and porosity for the adhesive. Lining paper reduces the possibility of bubbles and wrinkles by absorbing some of the moisture from the adhesive.

The wall should be prepared for the canvas by scraping off the flaking and peeling paint and spackling wide cracks or depressions (greater than $\frac{1}{16}$ inch in diameter). A wall-covering primer is then applied to the wall and the canvas affixed to the wall in strips like wallpaper.

If existing canvas has bubbled or come loose, there are several options.

■ The easiest solution is to readhere the loose canvas. The bubbled area is sliced with a razor blade, wallpaper paste is applied to the underside of the flaps, and the canvas is smoothed down.

■ If unsuccessful, the next choice is to cut out the loose piece and splice in new canvas. Finding canvas to match the

Partial removal of canvas from wall. New canvas will need to be spliced in to match the existing canvas. (Michael Devonshire)

148

existing material may not be easy. A piece of the existing canvas may be located in a hidden area of your building where it can be removed without notice.

- If replacement canvas cannot be found, the area can be resurfaced with joint compound. This usually will give the patch a different texture from the adjacent surface, however, so it should be done only as a last resort.

Existing canvas must be removed if readhering it will not work or if you see signs of plaster damage and need to assess the entire wall. Soft spots, bulges that respond to the touch or water stains are clues to underlying problems that require further investigation. Canvas often can be pulled off beginning with a loose corner. Other wall coverings usually are removable with chemical wallpaper remover, which is sprayed on small sections and allowed to soak in for several minutes before the paper is scraped off.

REPAIRING CRACKS

Plaster cracks are endemic in old buildings. Some cracks can be patched and never seen again. Others will recur no matter how often you try to eradicate them. You will probably find that your wall has a range of cracks from hairline to severe (wider than ⅟₁₆ inch). The treatments vary depending on the size of the crack, but most require the same tools and materials: a wall scraper, taping knives, all-purpose drywall joint compound and flexible mesh tape.

Hairline cracks are often cyclical. They will open and close as seasons change without threatening the stability of the wall. The best solution may be to leave them alone or, if there are many, to apply canvas to the wall or ceiling surface. They can also be filled with patching material.

Larger cracks tend to be stable in old buildings. If they are new or appear to be expanding, their cause should be determined by an architect or structural engineer before patching. Some cracks may require removal of adjacent unsound plaster and installation of new metal lath to reinforce the area.

For most large cracks, however, simple and small-scale repairs will usually suffice.

- The first step is to attach the adjacent solid plaster to the

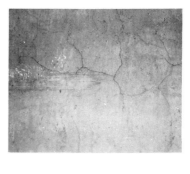

Top: Hairline cracks to be patched with joint compound or patching plaster or covered with canvas or wallpaper.

Above: Stress cracks running diagonally from door opening. These usually can be repaired with fiberglass mesh tape and joint compound. (both, National Park Service)

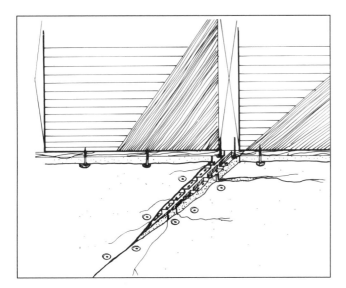

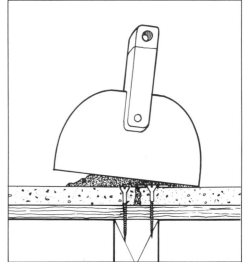

Left: Securing loose wood lath and plaster using plaster washers and lath nails to reattach lath to structural framing.

Right: Joint compound applied over plaster washers to fill crack. Blade is held at an angle to prevent ridges in patch. (*Old House Restoration*)

substrate with plaster washers if much movement will occur in the wall when the crack is opened.

■ The cracks are then undercut with a sharp pointed tool such as a putty knife or can opener to widen the area below the surface, providing a solid anchor for the patch.

■ Crumbs or debris are cleaned out.

■ The cracks are then filled with joint compound or quick-setting finish coat. Generally, the crack is filled with one coat of compound, covered with tape bedded in the compound, followed by another coat of compound.

■ The compound is feathered (spread with a knife to blend in with the level of the adjacent surface). Sometimes a third coat is required to smooth out depressions.

■ When dry, the patch is sanded and wiped with a damp sponge to remove residue and dust. It is then sealed with primer.

For gaps at the juncture of plaster and wood molding caused by humidity and temperature-related shrinkage and expansion of the wood, a highly flexible material such as latex caulk can be applied with a caulking gun. For other particularly stubborn cracks that refuse to be healed by ordinary methods, materials that can be used include a heavy-bodied high-adhesive compound or fiberglass tape and bedding material or synthetic resin adhesive, which are stronger and more flexible than joint compound and mesh tape.

When removing damaged plaster, contractors wear special masks to protect them against plaster and lead paint dust, a hat, long-sleeved coveralls and eye protection. Drop cloths should be spread on the floor because a good deal of plaster will end up there. Work should be done in a well-ventilated area, keeping temperatures between 55°F and 70°F.

Before beginning demolition, the adjacent solid plaster is secured with washers. The damaged plaster is cut out with a tool such as a chisel and gently removed with the help of a pry bar. Pounding on the wall should be avoided because it can loosen adjacent plaster.

Loose wood lath should be resecured to the framing. If missing or rotten, lath should be replaced with salvaged or new lath. Even if the existing wood lath is sound, professionals generally recommend attaching new metal lath over it for better keying and a stronger patch. Gypsum or rock lath nailed through the existing lath and into the studs also can be used. If the existing wood lath is exposed, however, it must be wetted thoroughly because it will absorb moisture from the new plaster. Before installing new lath, the broken plaster keys should be knocked off the old lath and vacuumed up. Depending on the lath, two or three plaster coats are applied, overlapping new and old plaster for stronger, more integral patches. Gypsum and rock lath require two coats of plaster; metal and wood lath require three.

Plaster with lime in it should cure at least one month before painting, while gypsum plaster should cure for two to three weeks. All patches and new plaster should be primed with alkaline-resistant paint specially formulated for priming new plaster.

Making a plaster patch. Left to right: Securing new metal lath to existing wood lath for better plaster keying; brushing water on edges of existing plaster to prevent it from absorbing all the moisture from the new plaster before it sets; applying the new plaster with a steel trowel. (John Leeke)

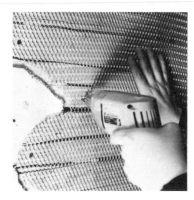
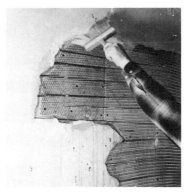

Where plaster is damaged or missing, holes can be patched with new plaster. Plaster patches are preferred over drywall to maintain the consistency and texture of the existing wall. Plaster can be applied to fit irregular holes and modulated in texture to match the adjacent surface. The straight lines and smooth flat surface of a drywall patch, however, are usually more visible. Also, since existing plaster is often not of uniform thickness, it is difficult to make a drywall patch flush on all sides.

For museum-quality restorations, historic plaster may be analyzed to determine exact proportions of particular ingredients required for the patching mixture. For rehabilitation work, it is possible to use available materials that match the historic plaster. Ready-mixed base-coat plaster consisting of gypsum and an aggregate such as perlite is generally acceptable for patching scratch and brown coats. High-gauge lime putty (half lime, half gypsum) also can be used to patch cracks and small holes or to repair a finish coat. The texture of the finish coat can be varied to match the existing surface by adjusting the size and amount of sand and by "floating" the surface with a tool such as a rubber-faced float for a rough surface or troweled with a steel trowel for a smooth one.

Larger holes — those more than ⅛ inch deep — or areas of damage that extend through the plaster to the lath usually require three-coat plaster patches. Often base coats of sand-aggregated gypsum plaster and a finish coat of lime putty and gauging plaster will provide a good patch. Perlite should not be used in the base coat for sizeable patches (more than 4 feet) because, in drying, the finish coat is subject to map cracking.

REPAIRING TEXTURED FINISHES

Textured finishes simulating stone, brick or ceramic tile were popular for a range of historic styles. Modern versions of these historic finishes may be purchased, usually in dry powder form to be mixed with water. They range in texture from smooth to sand finish or even coarser.

If the original textured finish was created in the plaster finish coat, a variety of plasters and techniques will re-create the effect. A colonial plaster finish, for instance, can be simulated with sand-texture paint applied relatively unevenly and then worked with a small pointing trowel and stippled with a brush. Imitation brick, stone and tile are made with colored plaster and wire, brushing the surface to achieve a rough texture. When the brown coat is a contrasting color, simulated mortar joints can be cut with a tool such as a screwdriver.

REPAIRING SCAGLIOLA

Scagliola can be repaired with plaster patches marbleized to match the existing surface. Commissioning replacement pieces of scagliola where large areas are damaged or missing is more problematic, as few craftspeople still practice this art.

DRYWALL PATCHES

Although drywall is an acceptable patch material for re-habilitation work, it is not recommended for small holes (less than 4 inches in diameter), because it requires the removal of a relatively large area of sound plaster to get to the lath or studs. If the ease and relative low cost of drywall make the difference between saving and patching a plaster wall or removing the plaster and installing a whole new gypsum drywall wall, however, go ahead and use drywall patches.

■ The first step is to remove the damaged plaster back to the studs, securing the edges of the adjacent solid plaster with plaster washers. (If the hole is small, the patch can be attached to the existing lath.)

■ The hole is squared up with tools, such as a framing square and utility knife or keyhole saw, and the plaster removed. The existing lath may remain in place but the sides of the hole should fall on studs.

- The drywall is then cut to fit the patch and installed with drywall screws. The patch may have to be doubled up or shimmed to make it flush with the adjacent plaster surface.

- After the patch is installed, joint compound is applied along the seams, followed by drywall tape bedded into the compound and another coat of joint compound.

- When dry, the seams are sanded and the patches primed before painting.

LOOSE PLASTER

Loose plaster may be caused by delamination of the finish coat from the brown coat, broken plaster keys or separation of the lath from the structure (see Evaluating Plaster for how to determine the source of loose plaster). If the problem is delamination of the finish coat, sometimes satisfactory repairs can be made by removing the loose plaster and patching the area.

Loose lath can be resecured to the framing with plaster washers and long drywall or flat-head screws.

NEW PLASTER KEYS

If the plaster keys are broken, several options are possible:
- Replastering the back of the existing plaster and creating new keys

- Using plaster washers to reattach the plaster to the lath

- Injecting adhesive epoxies

The first option may not be advisable for ceilings, however, as new plaster keys may add too much weight. On walls, though, if the back of the lath is accessible, the bowed plaster can be pushed against the lath from the front with padded boards and a brace, while the broken and loose keys are removed from the back. The old lath and plaster should then be wetted thoroughly and coated with a bonding agent. New metal lath is attached to the existing lath. A plaster mixture is then troweled into the spaces between the lath to form new keys. Reinforcement in the form of plaster-soaked strips of fabric can be added to the back of the lath.

INJECTED ADHESIVE BONDING

In some situations the back of the lath is not accessible or the lath is deficient, for instance, the spaces may not be large enough to allow adequate keys to support the plaster. You may even be faced with a plaster surface applied directly to the interior face of the exterior wood or masonry sheathing without any lath. Where the plaster must be preserved rather than removed and replaced, injections of adhesives to bond the plaster to the substrate may be necessary.

Injected adhesive bonding is particularly useful when there is a decorative finish that must be saved and the visible face cannot be disturbed. For restoration, conservators favor adhesives, because they can be injected through small holes, are strong, flexible and shrink resistant, and have low density, adding a minimum load (especially critical for a ceiling). They are also nonflammable and relatively nontoxic. Adhesive bonding, however, is not used as often in rehabilitation, because the materials are expensive and can be difficult to find, and the process requires the skills of an architectural conservator or an experienced restoration plasterer.

Typical steps include the following:

■ A conservator will first determine the areas of loose plaster by pressing lightly to see if the plaster moves separately from the lath, tapping the plaster and noting areas that sound hollow. A whole loose area should be adhered at the same time to minimize the stress on the plaster.

■ If the plaster has been damaged by water and is crumbly, it can be consolidated with an acrylic resin brushed and sprayed onto target areas. Consolidation is done at least a week before injecting the adhesive bonding.

■ An acrylic resin-based adhesive is injected through holes drilled in the back of the lath, if it is accessible. (Restoration plasterers may use more commonly available water- and solvent-based adhesives.)

■ If not accessible from behind, holes will be drilled through the face of the plaster.

■ Where possible, the broken keys should be removed and the debris vacuumed out from the area between the plaster and lath, as small chunks of plaster will get between the

Injecting adhesive between plaster and lath through holes drilled in the plaster. In ceilings (above), the ceiling must be braced in preparation for injected adhesive bonding. The bracing is kept in place until the adhesive between the plaster and lath has set. (John Leeke)

plaster and lath and prevent the plaster from being pushed back into place.

- If the wall or ceiling is bowed as well as loose, it must be pushed back into position and braced until the adhesive has set.

PLASTER SUBSTITUTES

If your plaster wall or ceiling is beyond repair and must be replaced, you have three choices: replastering the wall over the existing lath using traditional or machine applications; applying a veneer plaster to a gypsum board base; or installing drywall (removing all the lath must be considered here). Substitutes for traditional hand-applied three-coat plasterwork in old buildings should be used only when repair or replacement of plaster in kind is not feasible. It is important to remember that only hand-worked plaster looks like historic plaster. Even though veneer plaster may resemble plaster, it lacks the solidity and thermal and acoustical properties of plaster. Drywall, while it is efficient to install, has none of the character of plaster and also lacks its thermal and acoustical qualities. In addition, unless finished by an experienced installer, drywall joints are likely to show.

MACHINE APPLICATIONS

For large areas, or if plaster on an entire wall or ceiling needs to be replaced, plaster can be sprayed by machines through hoses onto the walls and ceilings. Windows, doors and other areas that require protection must be masked. The nozzle is held 18 to 24 inches from the surface and moved from side to side continuously to spread plaster uniformly. This work is usually done by a four-person crew — one person to spray, three people to follow and straighten up the work. Where metal lath is used, the lath is first "fogged in," or covered with a fine mist of plaster, so the plaster will not be blown through the openings in the lath. When this has partially set, the scratch coat is applied, followed by the brown coat. For a smooth finish coat, the plaster is usually applied by hand. Similarly, as with plastering by hand, screeds are established, the same number of coats are used, and the same hand-finishing techniques of rodding, darbying and trimming of the angles are required.

VENEER PLASTER

Sometimes called one-coat plaster, thin-coat plaster and rapid plaster, veneer plaster consists of one or two coats of specially processed high-strength plaster applied to "blueboard," a special gypsum-core board sandwiched between highly absorbent paper, or to standard rock lath or gypsum board, sprayed with a catalyst to accelerate the setting time of the plaster and create a strong bond. It can be applied by hand or machine. As with regular plaster, corner beads and edge terminals are used for the transitions between plaster and other surfaces. These will allow each material to expand and contract at its own rate. One-coat veneer plaster is faster, but two-coat plaster is stronger and smoother. The

Preparations for application of veneer plaster. Clockwise from left: Mixing plaster with a commercial rig; metal or plastic grounds installed at corners and reinforced with fiberglass mesh tape; transfer of finish coat plaster to the hawk for application over the scratch coat. (*Fine Homebuilding*)

157

Application of base coat of veneer plaster. Stilts enable plasterer to reach ceilings and high surfaces. (*Fine Homebuilding*)

surface can be troweled smooth, textured or tinted to approximate the historic plaster finish.

- The first step is to install the base (e.g., blueboard) and edge terminals.

- Corner beads are installed at the outside corners.

- Seams and inside corners are taped with fiberglass mesh bedded in veneer plaster.

- The plaster is then mixed with water and applied with a trowel in a thin $\frac{1}{16}$-inch layer.

- Using veneer plaster or a regular lime finish coat, a second coat may then be applied and troweled smooth.

Veneer plaster is extremely strong and durable. Also, it can be painted almost immediately. Another advantage for uninsulated masonry buildings is that insulation can be installed between the blueboard furring channels. The entire veneer system should equal the depth of the original plaster, thus maintaining the same relationship between the plaster surface and any ornamental features such as window and door moldings.

DRYWALL

Drywall, or wallboard, is made of gypsum compressed between layers of paper. The boards are manufactured in sheets 4 feet by 8 feet and $\frac{1}{4}$ inch, $\frac{3}{8}$ inch, $\frac{1}{2}$ inch or $\frac{5}{8}$ inch thick.

- Drywall is installed on the existing or new structure using nails or drywall screws, which are countersunk and the holes patched with joint compound.

- Seams are taped with drywall tape bedded in joint compound and coated with several more applications of joint compound.

- The joint compound is then sanded smooth when dry.

- Last, the surface is primed and painted.

Wallboard sometimes incorporates other ingredients to attain specific characteristics. For instance, asphalt is added to increase moisture resistance, while glass fibers, which bind together the gypsum during a fire, are used for fire-rated drywall.

ORNAMENTAL PLASTER TECHNIQUES

There are basically two types of plaster ornament: one, moldings run in place on a wall or ceiling or on a bench using a template and, two, cast ornament made in stationary molds and applied to a surface. Some plaster ornament may combine run moldings with embedded or applied cast ornament.

Plain plaster moldings without surface relief usually are run in place with a template. Ornament such as ceiling medallions, brackets and dentils generally are cast in molds off site, sometimes in more than one piece.

If you need to replace a missing piece of cornice, it can be cast in a mold, created directly on the wall (run in place), or made on a flat surface (called a bench) to be attached later to the wall. Most plasterers prefer to run small infill pieces on a bench and large sections in place. If run on a bench, large sections, especially if several are needed, may twist slightly and fit together badly. Also, if laid on a nonplanar surface before setting, large pieces can deform, making it even more difficult to fit them to the irregularities of the wall and ceiling.

Ornamental plaster cornice with modillions. (Michael Devonshire)

MOLDING RUN IN PLACE

Run-in-place ornament is the most difficult of all plasterwork. Professionals may take more than a week to make the cornice for one room. This work usually requires two people — one to mix and apply the material and one to run and clean the template.

To create a cornice:

■ A template, or mold, with the "negative" of the desired molding profile, is made from sheet metal and backed with a wooden handle called a "horse."

■ Temporary wooden tracks to guide the template are installed on the wall and ceiling.

■ For molding greater than 1 inch deep, the cornice is built out with blocks following the overall profile and covered with wire lath. (Deep ornamentation run in place is not made of solid plaster, because the plaster shrinks while setting and may crack.)

■ For molding less than l inch deep, a bonding agent is

159

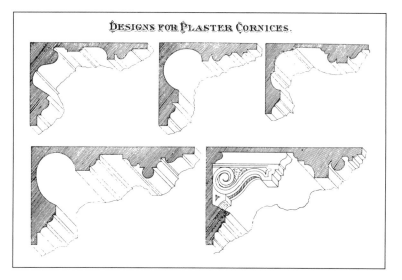

DESIGNS FOR PLASTER CORNICES.

Left: Template for molding cornice. The template has the reverse shape of the molding profile. (Michael Devonshire)

Right: Designs for plaster cornices from a late 19th-century pattern book, A. J. Bicknell's *Detail, Cottage and Constructive Architecture*.

painted onto the plaster surface to increase adhesion and to prevent the dry plaster from absorbing water from the new mixture.

■ Plaster is troweled onto the wire lath or wall or ceiling surface.

■ The template, or molding profile, is run across the soft plaster to form the profile.

■ More plaster is thrown on areas that need to be built up and to fill gaps and fine details and the template drawn across again.

■ Water is brushed on the surface to keep the plaster from drying too fast and as a final polish to create a smooth, shiny finish.

The template may be drawn across the surface many times to create a sharp profile. Plasterers must work quickly, as the cornice must be perfectly formed before the plaster sets.

Because the template can get within only a few inches of the corners, those areas must be molded by hand, with either a joint rod or a trowel, putty knife and sculpting tools; corners also may be run on a bench.

MOLDING RUN ON A BENCH

Infill and corner moldings are easier to run on a bench than in place. A molding is created this way:

- A parting agent such as stearic acid is brushed onto a flat surface to prevent the plaster from sticking.

- A line of plaster as wide and deep as the molding is then placed on the bench, and pieces of string or burlap are bedded in it for reinforcement.

- As with run-in-place molding, the template is then run over the plaster repeatedly, with more plaster added each time and hollows touched up.

- Toward the end, a new watery batch of plaster is mixed to make a milky film that can be brushed on to remove bubbles and irregularities and give the surface a polished appearance.

- Once the molding has dried, it can be shaped, planed, sawn, mitered, glued and nailed. When thoroughly set, the molding can be removed from the bench and installed on the wall or ceiling surface.

- Depending on the molding's size, it can be applied either directly to a plaster surface with a slip of gypsum plaster (sometimes mixed with glue), or it can be installed in a backing box and nailed or screwed into the lath, studs or joists.

When applying a molding directly to plaster, paint should first be stripped from the wall and ceiling. If a gypsum plaster slip is used to adhere the molding, the back of the molding is scored to improve the bond and the surfaces of the molding dampened to prevent them from soaking up

Molding run on a bench. Left to right: Template drawn across wet plaster to shape it; depressions filled in with more plaster to create an even surface; template drawn across plaster repeatedly. (Michael Devonshire)

161

Creating plaster ceiling medallion with mold mounted at center. (Martin Weil)

the water of the slip. Slip then is applied to the wall, and the molding held in place until it dries. The molding can also be nailed in place with finish nails in predrilled holes. The nails are set and the depressions filled with plaster or spackle.

When a backing box is used, molding is nailed into the wood box with finishing nails. Plaster first must be removed from the ceiling and wall area where the molding will be installed. The gaps between the infill and original molding can be filled with gypsum plaster and trimmed with tools such as a joint rod and straight edge.

CAST PLASTER ORNAMENT

Cast plaster can be used to reproduce detailed parts of or entire ceiling rosettes, moldings, decorative plasterwork and even wood carvings. Many companies offer prefabricated plaster or plasterlike ornament, and others will custom-make cast plaster ornament to specifications.

Castings must be made using molds created from existing pieces of ornament. If an original still exists and must be matched, the process is relatively easy. If none of the ornament you need to replicate has survived, an accurate wood, clay or plaster model must be created from which a mold can be made. It is easiest if the piece to be replicated or the model is detached. If removing the model from its place on the wall or ceiling would damage it, however, molds can be created in place. Ceiling rosettes in particular may be difficult to take down, although they sometimes can be disassembled in place or removed with the surrounding ceiling plaster.

Making molds

The choice of materials for molds depends on the cost, the complexity of the model and the number of castings required. Whichever molding compound is used,

■ The model usually is stripped of all paint and finishes; holes and cracks are repaired; and the edges are sealed with modeling clay.

■ Several coats of shellac are applied to seal in any impurities.

■ If necessary, the model is then painted with a mold-release agent, such as neutral soap or Vaseline, which makes it easier to separate the mold from the model.

■ The model can be put in a box, or wood or sheet-metal walls can be built around it to contain the liquid molding compound.

■ The molding compound — urethane is one of the most common — is brushed or poured over the model and allowed to cure.

■ The object is recoated, and layers of cheesecloth or burlap are sometimes inserted in between coats for reinforcement and to reduce the amount of molding compound required. Most molds ultimately are about ⅛ to ¼ inch thick.

Making plaster casts

Once the mold is ready, the ornament can be cast. The choice of casting materials depends on the desired surface finish, weight and strength. The traditional and most common casting materials are casting and molding plasters made from gypsum plaster. These may be heavy for large pieces, however, and other materials, such as latex casting

Filling a rubberlike mold with plaster to cast a decorative cornice (top) and removing the rubber mold from original plasterwork. (Balthazar Korab)

163

rubber, polyurethane foam, fiberglass and plastic resin, can be used to achieve finer detail or different surface finishes.

Small plaster casts are made by pouring a plaster compound into the mold, jiggling and tapping it to fill the crevices and bring air bubbles to the surface. The mold is usually supported in a form, or mother mold, which prevents the mold from distorting when the casting material is poured. The exposed surface is then leveled and, when it begins to set, scratched to help it bond to the wall or ceiling when installed.

Large castings are made so that they are hollow to reduce their weight. After the plaster mixture is poured in, the sides are built up with a material such as burlap. Deep or complex casts are reinforced with string, wire, burlap or hessian, rough-weave cloth of jute yarn. Wood lath, soaked in water first so as not to absorb moisture from the plaster, can also be added to reinforce plaster and provide a means of attachment to walls or ceilings.

When the plaster has cured, the mold is peeled off, the excess trimmed and any bubble defects repaired. If the cast was made in pieces, they are assembled with small nails using predrilled holes and set in adhesive.

Installing cast plaster ornament

Cast ornament is installed in several ways. Light moldings can be attached with a thin coat of plaster of paris or with mastic adhesive, epoxy or gypsum board joint cement. The older plaster surface is brushed first with a commercial bonding agent to increase adhesion. Heavier castings require screws. Once installed, the screw holes and seams are spackled and sanded.

Plaster molding substitutes

In situations where you have adequate documentation to create a matching replacement and a plasterer's skills are not available but carpentry skills are, you may be able to use built-up or milled wood moldings to replace historic plaster ornament. The wood can be given a plasterlike finish with a coat or two of acrylic gesso, a mixture of plaster, white lead and oil used by artists to seal canvas and by woodworkers to conceal rough wood surfaces and poor-quality joinery. Flexible latex caulk should be used to fill the gaps between the wood molding and adjacent plaster

Plaster castings and wood models from which the castings have been made. (Balthazar Korab)

164

surfaces, because wood and plaster have different coefficients of expansion.

A number of companies also manufacture ornament that resembles plaster, such as ceiling rosettes and tiles, cornices, brackets and column capitals. These are made from synthetic or plaster mixtures and are usually lightweight, easy to install and relatively inexpensive compared to the cost of hiring a plasterer to replicate ornament. This over-the-counter ornament should only be considered when there is documentation to replicate accurately the historic ornament or the replacement is appropriate for the age, style and class of building.

Fire codes may restrict the use of ornament made from petrochemically based materials, such as polymer, which release toxic gases when burned. (Check manufacturer's data and applicable building codes before installing plaster ornament of this type.) Products made of plaster with fiberglass matting, which reduces the weight and adds strength, are also available and have fire-resistant qualities. Manufacturers can supply fire-resistant data for any product you are considering.

Reproductions of historic plaster cornices. (Focal Point, Inc.)

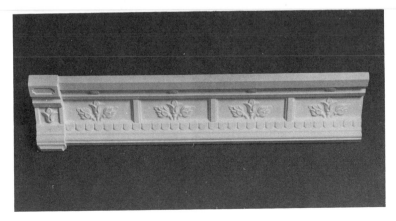

165

REVIVING DECORATIVE PAINTED FINISHES

Painted finishes can suffer equally from wear and tear or from misguided maintenance. Painted surfaces are all too easily marred by furniture scrapes; fingerprints and airborne dirt; holes for picture hooks, wall sconces and built-in furniture; discoloration and cracking from exposure to sunlight and heat; peeling and chalking from dampness and leaks. A certain amount of discoloration or degradation also may result from aging of the finish itself. Plain painted finishes may suffer further from the buildup of too many coats, obscuring original ornamental detail, or the interlayering of oil and water-based paints, resulting in cracking or peeling.

Typical solutions to these problems — intended as "maintenance" measures — often exacerbate the original damage: cleaning with an inappropriate cleaner that abrades or removes some of the paint; painting in damaged areas with the wrong color or type of paint, resulting in obvious, distracting touch-ups; applying a "protective" coating that discolors; or, most drastic of all, hiding the decorative finish completely with a new coat of paint.

Most damage to old and historic paint can be cleaned and repaired, with localized touch-ups to areas of missing paint. However, before you consider how to treat your painted finish, make sure that the surface underneath it is in good condition. Problems such as loose plaster or buckled veneer should be addressed before attending to the finish. Rehabilitating painted finishes should be one of the last steps in your preservation project.

Damaged hand-painted finish requiring careful consolidation and inpainting by an art conservator or restoration painter. (Jack Boucher)

Opposite: Victorian interior with stylized stenciled frieze and ceiling borders. (Library of Congress)

Beginning with the gentlest means possible, most surfaces can be rejuvenated by sensitive cleaning to remove dirt, grease, and superficial stains and scratches. Once the surface has been cleaned, you will probably find the remaining miscellaneous scuffs and scrapes acceptable. The objective is not to make the finish look like new, but to reveal the original pattern and character. Marks of age should be considered part of the finish's character — testimony to its history.

You can do most of the work of rehabilitating an existing plain painted finish yourself (see Rehabilitating Existing Finishes in this chapter), or you may choose to hire a painting contractor. Decorative painted finishes, however, are trickier. They generally are more fragile and may have been painted or coated with a substance that does not respond well to soap and water. They also may require delicate repairs or painting of severely damaged areas. Here the services of a painting or architectural conservator or a restoration painter are in order.

You will also need a professional conservator if you want to restore a decorative finish that has been covered with opaque paint. If, after removing the paint, the original finish below is too damaged to be restored, the original pattern and colors must be documented in order to reproduce them accurately. Where the overpaint cannot be removed to reveal the original pattern, you may need to paint the walls a historically appropriate or "period" color instead (see Choosing an Appropriate Finish in this chapter). Choose a color and finish (gloss, semigloss or flat) appropriate to the period and character of the room. Paint samples can be analyzed to determine the original paint color (see Paint Analysis in this chapter), or, if that is not feasible, a restoration architect, architectural historian or conservator can help you plan appropriate finishes.

A number of paint manufacturers have developed palettes of paints based on or adapted from historic colors, and several companies produce milk and calcimine paints. The use of modern materials and pigments, however, prevents exact replication of the color, texture and finish of their historic prototypes. It is always preferable to save the

Los Angeles Central Library. Here, the soiled and stained stenciling can be cleaned by an art conservator, the fire extinguisher removed and placed where it will not mar the building's historic fabric, the door and baseboard cleaned and repainted. Better and more harmonious signage would also play a part in this rehabilitation. (©1990 Bob Ware)

TYPES OF DECORATIVE FINISHES

Stenciling: Repeated patterns of painted decoration applied with templates and brushes or sponges. Popular from the 17th through the early 20th century.

Glazing: A film of transparent color applied over an opaque paint color to enrich the color or to achieve a richly textured surface. Popular particularly in the 1920s and 1930s. Includes the following processes:

■ Mottling or "sponging": A random mottled texture achieved by blotting the surface with rags, sponges or newspapers.

■ Striating: Short vertical stripes created by dragging a tool such as a brush or steel wool through the glaze.

■ Stippling: A dotted or pebbled finish created with a stipple or dry paintbrush.

■ Rubbing: A parchmentlike effect achieved by "rubbing out" the glaze.

■ Polychrome: Several colors of glaze, with one predominant and others for highlights.

■ Graining: Painted imitation of wood, often a finer and more distinctive species than the wood of the actual surface; also used on nonwood surfaces such as plaster. Often done on doors, window sash, baseboards, paneling and fireplace mantels and surrounds.

■ Marbleizing: Painted imitation of marble, used to dress up or disguise more ordinary surfaces. Often done on door frames and paneling, baseboards, fireplace mantels and surrounds, and floors.

Gilding: Gold leaf or other products such as metal leaf applied to simulate solid gold. Particularly popular for neoclassical styles.

Top to bottom: Graining, marbleizing and stenciling. (Bill van Calsem; Michael Devonshire; Thomas Sweeney)

169

original finish where it is salvageable, because no modern substitute can accurately reproduce all its nuances or its patina of age.

REHABILITATING EXISTING FINISHES

Rehabilitating plain and decorative historic painted finishes — reviving and repairing them — encompasses a range of activities that vary for plain or decorative painted finishes.

PLAIN PAINTED FINISHES

If the existing paint finish is stable and only superficially cracked or peeling, it probably can be rehabilitated or re-painted without wholesale removal. Washing will revive a finish that you want to retain or, if it is necessary to paint over it, will allow a good bond with the next layer. Here are some basic techniques:

■ Vacuum or brush off dust and fibers with a soft brush or feather duster. Vacuuming will lift the fibers and is preferable to using a rag, which will smear them instead. (It is important to maintain painted finishes by dusting them regularly to prevent dust and fibers from becoming embedded in surface oil or grease.)

■ Wash with water and a gentle detergent to remove oil and grease. (A discreet patch should be tested for water solubility or other reactions first.) Cleaning can be done using a sponge and mild detergent solution, rinsing the surface thoroughly when the dirt has loosened and drying it with paper towels.

■ Use mineral spirits, followed by soap and water, to remove heavy layers of grease, which may be found on kitchen walls and ceilings.

■ Once the surface is clean, touch up disfiguring scratches and nicks with leftover paint or artists' acrylics.

If you find that repainting is necessary even after cleaning, the next step is to remove cracked, loose or flaking paint with a scraper or wire brush. Any damage to the plaster or wood substrate should be patched and primed. The painted surface can then be sanded, wiped clean and repainted to match the existing or historically appropriate color (see Paint Analysis for information on determining original paint colors).

Casein and milk paints

Casein or milk paint generally can be removed with commercially available full-strength ammonia and fine steel wool, although the ammonia may have to remain on the paint for 10 minutes or more to dissolve it. Wear rubber gloves and make sure the area is well ventilated. Ammonia is highly caustic and toxic, and minimal precautions include use of a respirator mask with a vapor filter.

Calcimine paint

Calcimine paint will not bond with other paints, so it must be removed before repainting. A mixture of trisodium phosphate and hot water, scrubbed on the surface with bristle brushes, is usually effective. If calcimine paint lies under later layers of paint, it may have caused them to peel as well, in which case all the paint must be removed. Heat and chemicals are typical methods for removing paint (see Methods of Stripping in the chapter on woodwork). If conventional heat and chemical methods are used to remove the upper layers, the calcimine paint must still be washed off once exposed. An alternative that often works where calcimine is present is the application of steam, which will loosen the calcimine layer, thus allowing all the top layers to be removed with it.

DECORATIVE PAINTED FINISHES

If a decorative painted finish is covered with layers of dirt and grease, it usually can be cleaned. You may try doing this yourself, although most experts recommend hiring a professional because the use of inappropriate materials or methods can cause irreversible damage. Conservators or restoration painters begin with small test patches in a hidden area, trying a series of cleaning solutions. They generally begin with the mildest solution and gradually work up to stronger ones until they find one that will clean the surface without damaging the paint. The types of paints and coat-

Cleaning a decorative finish. Testing a cleaning agent on stenciling (top) to determine the most effective and least harmful to the historic paint and a light patch (above) showing the dramatic difference after a deteriorated coating has been removed. (Natalie Shivers)

ings used in the finish will determine the best cleaning methods and materials. Variations in humidity and types of dirt will also influence the choice of treatment.

Conventional cleaning techniques

You may try several conventional cleaning solutions on test patches, watching carefully to make sure that you are removing only dirt:

■ Dust the surface with a soft brush and, using a soft rag or sponge, gently wipe with a highly diluted (2 to 5 percent) solution of detergent (such as trisodium phosphate or ammoniated cleaner) and water.

■ Rinse the surface well several times to remove any residue.

■ If the paint layer seems to come off with the dirt, the paint is water soluble, so you should try mineral spirits instead.

■ If either of these solutions is effective, it may be usable on the rest of the surface. Be careful, however, as some colors on the same surface may respond differently to the cleaning treatment. If that seems to be happening, stop cleaning and call in a professional.

Special solutions

You will also need the services of a conservator or restoration painter if conventional cleaning methods do not work. Conservators may use dry methods such as erasers or "scum bags" (filled with an erasing powder) on distemper (water-based) paints, for instance. If these techniques are not effective, chemicals such as mineral spirits, naphtha or xylene may be used. Oil paints also may sometimes be cleaned with dry methods. Alternatively, conservators use solutions of various types and strengths of detergents and water, mineral spirits or emulsions combining the two. Occasionally, different solutions work on different colors and may require more than one washing.

Protective coatings

If a protective coating such as shellac or varnish has discolored or deteriorated, it may be dissolved using cotton balls with a solvent, such as mineral spirits for varnish or denatured alcohol for shellac.

Once cleaned, the condition of the paint surface can be

evaluated. While the objective is not to make a decorative finish look new, your conservator may paint in disfiguring scratches and areas of missing paint. Conservators generally apply a sealer such as varnish between the original and new paint, so the new paint can be removed without damage to the earlier paint surface.

Removing overpaint

Overpainted images can sometimes be uncovered and restored with careful conservation techniques. Overpaint removal should be done by a professional conservator as it requires knowledge of many alternative chemicals and methods in order to find one that will remove the overpaint without removing the decorative paint. It is essential to do solubility testing to determine if the overpaint is water based, oil based or another type such as casein paint.

Gilding

Preserving gilded surfaces presents special problems, some of which arise from the original choice of gilding materials.

Pure gold leaf does not fade, discolor or tarnish, nor does it flake or peel if applied properly because it is elastic and flexes with the movement of the substrate. It will require periodic cleaning of superficial dirt, however. This must be done with great care by a gilding conservator, as it is difficult to remove the dirt without damaging the patina. Depending on the type of gilding base (water or oil), mild detergents or various combinations of chemicals may be used. Water gilding, for instance, can sometimes be cleaned with a mixture of acetone and xylene, while oil gilding may respond well to water-based solutions such as dilute ammonia. A conservator will begin with the mildest solutions and try progressively stronger ones until a successful formula is found.

Coatings such as glue or shellac may discolor or degrade and require cleaning or removal. Removing the coating without removing the gold leaf is a delicate process and may not be successful where finishes have chemically linked together. These areas may require regilding.

Metal leaf, a popular form of which was a copper alloy known as Dutch gold or Dutch metal, will discolor, turning green over time, if it was not coated with a protective coating such as varnish or shellac. Once discolored, the surface

Heinz Hall, Pittsburgh, formerly a movie theater. Gilding here highlights the ornamental features. (NTHP)

Diagnosing Painted Finish Problems

Plain Painted Finishes

Damage	Cause	Solution
Dirty, grimy surface	Pollutants, kitchen grease, body oils, general use	Clean surface with detergent or chemicals, such as mineral spirits.
Wrinkled paint	Top layer painted before earlier one dried; excessive amount of paint applied at one time; temperature too hot or cold when painted	Sand or scrape off wrinkled layers(s); repaint.
Loose, flaking or peeling paint	Moisture, unclean surfaces under paint, paint applied in direct sun	Scrape, sand or wire-brush loose paint to reach sound layer; repaint.
	Calcimine paint	Wash off with hot water and trisodium phosphate.
	Calcimine paint under layers of other paint	Remove later paints with heat, chemicals or steam; wash off calcimine paint with hot water and trisodium phosphate.
Cracked, crazed paint	Aged, inflexible paint film	Scrape, sand or wire-brush paint to reach sound layer; repaint.
Alligatored paint	Interlayering of incompatible paints or finishes; inflexible paint film	Scrape or sand off paint to reach sound layer; repaint.

DECORATIVE PAINTED FINISHES

Problem	Cause	Solution
Dirty, grimy surface	Pollutants, kitchen grease, body oils, general use	Clean surface with detergents or chemicals.
Discolored or degraded coating	Age, heat, sun	Remove coating without damaging decorative finish; apply new protective coating.
Flaking, peeling paint	Age, heat sun, moisture, incompatibility with other paints or finishes	Remove loose paint; paint in decoration; apply new protective coating.
Overpainted finish	Paint applied over original design	Remove overpaint and restore original finish; or document and re-create original finish; or if the original finish cannot be documented, apply a historically appropriate finish.

Typical problems of painted finishes. Soiled decorative painted surface, which should be cleaned by a professional (far left, NTHP), and dirt and grime on a plain painted finish (left, Julian Graham).

must be regilded. Even if the leaf is still in good condition, but the finish has discolored or degraded, it is often difficult to remove the finish without removing the metal leaf. In both situations, the surface should be cleaned and removal of the degraded finish tested. If those attempts at rehabilitation of the existing metal leaf are not successful, the surface should be regilded. Because the same discoloration or degradation will recur with Dutch metal, gold leaf should be used if regilding is required.

Bronze paint made with bronzing powder mixed in various solutions tends to oxidize, turning iridescent and then brown. It, too, usually requires removal and replacement when degraded. Roman gilding, or burnishing bronze consisting of bronze powder suspended in a glue solution and brushed on a surface, is more stable and may require only superficial cleaning.

Overpainted gilding of all types is difficult, if not impossible, to restore. If testing proves this to be the case, you should consider regilding the surface.

CHOOSING AN APPROPRIATE FINISH

If the historic stenciling, glazing, graining, marbleizing or gilding on your walls, moldings, woodwork or ceiling is damaged beyond repair, or if overpaint cannot be removed without hurting the original decorative paint, the finish can be re-created from the pattern and colors of the remaining finish.

If only vague traces of the original decorative finish remain, however, you should consult with an architectural or painting conservator, architectural historian or restoration architect who can supplement evidence found in the building with knowledge of decorative finishes popular for the period. Such a professional also will look at treatments in other buildings of the same type and age in your area to determine an appropriate color and pattern. This local research is important because the colors, types of wood or marble imitated and realism of the effect can vary widely, depending on the skills of the painter as well as prevailing taste. Stencil patterns and colors also reflected current styles and available skills.

In all cases, it is important that any work done is reversible and can be removed without damaging the finishes underneath. If possible, all the existing finishes should be left intact and any new work added as a top, reversible layer with an isolating layer of sealer applied between old and new paint. If the existing finishes must be removed because their condition is too deteriorated to provide a stable base for a new finish, it is important to leave at least a representative patch of all finishes on each surface to maintain a record of the history of the building.

DOCUMENTARY AND PHYSICAL RESEARCH

The existence of decorative finishes — stenciling, glazing, graining, marbleizing or gilding — that have been painted over can be determined by paint analysis, but written and historical documentation may provide the first clues to their existence.

Documentary evidence includes artists' and workmen's bills, photographs, descriptions in building committee records or letters. Decorative painting in other buildings of the same age and type in your area or advertisements in local newspapers also may suggest the presence of a decorative painter who could have worked on your building as well.

Physical evidence of earlier decorative paint sometimes can be found in the building itself by shining a raking light across walls and ceilings. Shadows created by variations in the paint film surface or thickness may be visible.

One of the first steps a painting restoration specialist may take is to research and document the locations of the finish. If the original finish was painted over, numerous paint scrapings may be required on moldings and doors and in the corners of walls and ceilings, as well as along their borders and fields.

Once the original finishes are uncovered, it may be possible to expose a portion if removal of all overpaint is not feasible. The colors can then be analyzed and the pattern documented with photographs and full-scale drawings. Because the patterns and colors may vary from wall to ceiling and from corner to border to field, documentation of all portions should be completed to represent the finishes in all locations.

Top: Original stencil painted with varnish glazes from the Cornish House, Little Rock, Ark., revealed after repeated applications of paint remover to break through years of accumulated paint layers. The deterioration of the original pattern led to a re-creation of the stenciling.

Above: Re-created stencil. (both, Rebecca Witsell, Studio Werk)

Opposite from top left: Damaged and re-created ceiling stenciling; drawing documenting the original stenciling; new stencil based on the original design. (Pinson & Ware)

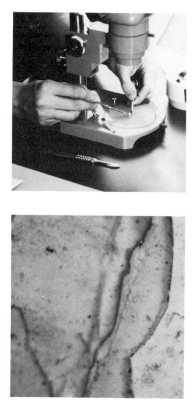

Paint analysis. Examining a paint sample under a microscope to determine the sequence, types and colors of finishes (top) and a microscopic view of paint layers. (Martin Weil)

Analysis of original paint colors is essential to re-create plain and decorative painted finishes accurately and is a valuable tool in restorations. For rehabilitation, however, you may need to postpone it while you take care of essential patching and repairs because of the time and expense involved. If that is the case, make sure to retain sample areas of all paint layers on all walls and moldings for future documentation.

Professional paint analysts

Although you may be able to scrape down to the original paint layer yourself and may be tempted to draw conclusions about colors, only a professional paint analyst can accurately assess the results. Using microscopical analysis and microchemical and ultraviolet bleaching techniques, the analyst can determine the number and nature of paint layers, original colors, distribution of colors within a room, existence and appearance of decorative painted finishes and approximate date of each layer. A professional will be able to determine the difference between a prime and a finish coat, for instance, and to compensate for changes in color resulting from aging, sunlight, dirt and oils.

Paint analysts take sample cross sections of all paint layers from a variety of surfaces such as walls, ceilings, doors, windows, moldings, paneling and mantelpieces. The samples are then examined under a microscope and sometimes treated with chemicals or bleached with ultraviolet light to determine the original color. Colors are matched with the Munsell Color System, a standardized system used by all architectural and paint conservators for cataloging paint colors. Their hue, value and chroma are assigned letters and numbers that correspond to Munsell color chips. Commercial paints can then be mixed to match those chips.

Self-sampling

If the cost of hiring a professional paint analyst is not currently in your budget, it is possible to take the samples of paint layers yourself. You can send them to a paint specialist for instrumental laboratory analysis and color matching. However, be sure to retain samples of original paint

layers in all the rooms, on walls as well as moldings, because your samples may not be truly representative. For example, they may not have all the paint layers or they may be of poor quality for matching colors.

REPLACEMENT FINISHES

Surfaces to receive re-created decorative finishes are generally cleaned and, where traces of the original finish remain, sealed with a protective coating such as varnish or shellac and then patched and painted. Experts sometimes recommend applying canvas to a wall or ceiling to be stenciled. This can even out rough sections, bridge hairline cracks and provide a surface for decoration that will not display minor plaster cracks (see the chapter Rehabilitating Plasterwork).

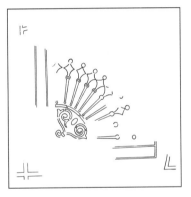

Stenciling

Once the original layout, patterns and colors of stenciling have been determined, a restoration painter will draw plans to scale, laying out the stenciling on each surface and figuring the spacing of motifs. The contractor will then make hard-line drawings from which the stencils are produced. A pattern may require several stencils, one for each color. Stencils can be made from a variety of materials, although acetate or mylar is generally favored. The patterns, traced onto the stencil material and cut out with a utility

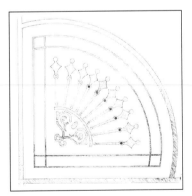

Artisan brushing paint through a stencil applied to a ceiling. (Pinson & Ware)

knife, are made with "bridges" or pieces that hold the pattern's form together. Stencil paints range from japan paints to flat, oil-based paints and are applied with round brushes or small pieces of sponge.

Patterns are laid out on walls and ceilings using rules, plumb bobs and chalk lines. The actual stenciling is done by holding the stencil tightly against the surface (to prevent paint from leaking behind it) and applying the paint with a relatively dry brush in a swirling motion. The stencil is then lifted up and moved to the next location. So as not to cause smearing, each stencil must be dry before the next color can be applied. When completed, the bridges holding the stencil together are painted in and the surface is given a protective clear coating.

Glazing

Restoration painters use a variety of techniques and materials to re-create glazing after documenting the original location and type of glazing (colors, pattern, texture). Usually the process is as follows:

- A nonporous base coat such as a semigloss alkyd or high-gloss latex paint is applied on a clean, stable surface.
- When the base coat is dry, the glaze is applied over it.
- Glazing is done usually in teams of two people, one to apply the glaze, the other to texture it. They will probably do one whole surface at a time, maintaining a wet edge to avoid lap marks.
- The glaze is allowed to dry thoroughly before varnish is applied for protection.

While specialists often use their own mixtures for glaze coats, glazing can also be done with a tinted commercial glaze. Glaze coats usually consist of various combinations of varnish, linseed oil and turpentine or mineral spirits, sometimes with paint and tinting colors added. The color, texture, opacity and consistency of the glaze can all be varied to achieve a range of effects.

Four basic glazing techniques were used historically and are still practiced today:

"Sponging off." To achieve a mottled or softly marbled effect, the wet glaze is "pounced" or dabbed in a random pattern with a wad of cheesecloth, tissue paper or sponge, rotating the cloth and rearranging the folds frequently to

Glazing door panels and trim by "sponging off." (Rebecca Witsell, Studio Werk)

avoid repeating the pattern. Another popular technique uses a rag or sponge dipped in mineral spirits and blotted on newspapers before pouncing the glaze. The sponge lifts some of the glaze while the mineral spirits soften and blend the remainder.

"Sponging on." With this process, a variation of the "sponging off" technique, a very thin glaze is applied to the wall with a natural sponge to achieve a crisper, more distinct texture. The sponge is dipped first in mineral spirits, then in the glaze, and blotted on newspaper before being applied over the base coat. Other glaze colors can be applied when the first coat is dry.

Striating. The process of striating consists of dragging a dry brush (or rag, sponge or steel wool) through the wet glaze in short parallel vertical lines.

Stippling. For a stippled finish, the glaze is given a dotted or pebbled finish with a dry brush such as a stipple brush. A fluffy paint roller, wad of cloth, newspaper or sponge also can be used.

Graining

Professional decorative painters have their own special methods and formulas for graining. Basically, however, graining can be done by either a "positive" or a "negative" method, in other words, by adding or removing paint. The graining glaze is traditionally a combination of

Graining tools from a 19th-century illustration. Graining comb, lining tool, badger softener, veining fitch. (The Athenaeum of Philadelphia)

181

Top: Graining door panel.
(Jack Boucher)

Bottom: Marbleizing base-
board. A brush and feather
are being used here.
(Thomas Sweeney)

boiled linseed oil, varnish, paint thinner, paint drier and tinting colors, although commercial glazing liquids also are available.

One common method of graining includes these steps:

■ A base coat, tinted to match the lightest shade of the wood to be imitated, is applied first and sanded when dry.

■ A darker glaze is brushed on over the dry base coat.

■ A rubber comb or stiff bristle brush is then used to remove the darker glaze in irregular, wavy striations, which may be broken up by stippling with dry brushes.

■ Areas can be lightened with cheesecloth or a sponge, while a chamois tip, piece of cork or blotting paper can be used to create knots, which are then outlined and highlighted with paint. Veins can also be painted in.

■ The entire surface may then be whisked with a badger blender to soften the edges.

■ A third tinted glaze coat may be applied when the surface is dry.

Marbleizing

Marbleizing consists of a tinted glaze brushed over a base coat, sponged lightly and veined with paint. It can also be done by removing the wet glaze to reveal portions of the base coat, a method usually used for imitations of dark marble.

The following is a typical marbleizing method:

■ A ground color matching the predominant color of the marble to be imitated is applied with a brush, sponge or cheesecloth.

■ When the ground color is dry, a coat of one or two colors is applied and mottled with a sponge, wooden craft stick or waddle to create a marbleized look.

■ Veins are made with a feather or thin brush dipped in mineral spirits and then in the color and run over the still-wet ground.

■ Areas may also be blended and manipulated with tools — the point of a feather dipped in mineral spirits or a dry blender brush to alter the pattern; or a sponge, crumpled tissue, newspaper or cheesecloth to create different textures.

■ A protective coating is applied as the final step.

Water gilding process. Top, left to right: Gilder's tools — gilding cushion, knife, tip, mop and book of gold leaf; bole brushed onto gessoed surface; gold leaf applied to surface.
Left: Tip pulled away quickly after depositing gold leaf; burnishing leaf on surface. (Stanley Robertson)

Gilding

The technique of gilding is painstaking and complex, requiring the sophisticated skills of a professional gilder or restoration painter to do the following:

■ The surface to be regilded is patched and cleaned so that it is stable and smooth.

■ The surface is then coated with a mixture of glue, chalk and bole (gilder's clay).

■ A water- or oil-based gilding technique is selected. Water gilding uses a solution of alcohol and water as a base for the leaf and produces a more brilliant surface, while oil gilding uses a varnish called size and produces a soft luster.

■ The gold leaf is applied to the surface leaf by leaf and smoothed with a gilder's tip, a thin flat brush of badger or camel's hair between pieces of cardboard.

■ When the entire surface has been gilded, it is burnished with an agate-headed tool and then coated with a layer of shellac.

183

GLOSSARY

Acanthus Plant whose stylized leaves decorate the Corinthian capital; also used as a motif on other parts of a building such as friezes and panels.

Adobe Sun-dried clay that has been mixed with sand and sometimes gravel, straw or grass; used as a construction or finish material.

Anthemion Greek ornament derived from the honeysuckle or palmette; used singly or in bands.

Architrave Lowest component of the three primary divisions of the entablature in a classical order; also, molding around window and door openings.

Astragal Also known as a bead; a convex, semicircular molding; one of eight classical molding shapes.

Back Band Outer molding of door or window casing.

Baseboard Also known as skirting; molding at the juncture of the floor and wall, sometimes divided into three parts: the base mold or cap, the base and the shoe.

Batten Board, narrow or wide, nailed on the back or face of two other boards to hold them together; also a molding used to conceal the line where two parallel boards or panels meet.

Bead Also known as an astragal; a small, rounded molding; one of eight classical molding shapes.

Bed Molding Molding beneath the corona and above the frieze; also refers to any molding beneath a projection, such as a cornice, or the lowest of a band of molding.

Board and Batten Type of wood paneling or exterior wood cladding consisting of flat boards joined with battens, usually in a vertical position.

Bole Type of clay used as a base for gilding.

Bolection Heavy projecting molding used to cover the joint of two elements often in different planes.

Calcimine Paint Also spelled "kalsomine" in the 19th century; a distemper paint made of tempera colors, water and sizing.

Canvas Fabriclike covering used to hide minor defects in walls or ceilings and to serve as a stable substrate for paint.

Capital Topmost member, usually decorated, of a column or pilaster; each of the five orders had its own distinctive capital.

Carton Pierre Form of composition plaster made of powdered chalk and a binder; manufactured in 18th-century England and used by Robert Adam.

Casein Paint Also known as milk paint; paint made with a binder of milk.

Casing Molded or flat visible trim or framing around a door or window opening.

Cast Plaster Plaster ornament formed in molds.

Caul Flat metal or wood plate used to protect surfaces and apply even pressure during repairs such as gluing.

Cavetto Also known as a cove; hollow molding with the profile of a quarter circle; one of eight classical molding shapes.

Chair Rail Horizontal molding at the height of chair backs to protect walls from being scraped; used alone or applied to the tops of dadoes or wainscots.

Chamfered Molding Molding with a beveled edge.

Composite Order Most elaborate of the five classical orders, combining elements of the Ionic and Corinthian orders; developed by the Romans.

Composition Ornament Plaster ornament made of whiting, glue and linseed oil and formed in molds.

Console Bracket in the form of an S-shaped scroll.

Corinthian Order One of the five classical orders; developed in Greece 5th century B.C. and conventionalized by the Romans. The design of the capital is said to have been inspired by the image of a basket overgrown with wild acanthus placed on the grave of a Corinthian girl.

Cornice Crowning or uppermost component of the three main divisions of the entablature; also used to describe any horizontal molding forming a main decorative feature, especially the molding at the intersection of the wall and ceiling.

Corona Overhanging element of the cornice, above the bed molding

Crosette Horizontal extension of the architrave molding at the upper corners on such elements as doors, windows and mantels to form "ears."

Crown Molding Molding serving as the corona or crowning member of a structure; most often used at the juncture of wall and ceiling.

Cymatium Also known as a cyma recta, ogee or talon; an S-shaped molding; one of eight classical molding shapes.

Dado (die) Portion of the pedestal of a classical column between its base and surbase; a term also applied to the lower portion of a wall when decorated separately or defined by a dado or chair rail.

Darby Tool, usually equipped with two handles, used to compact and straighten plaster.

Daub Mixture of clay and hay used as filling between vertical members in timber-framed houses; wattle and daub was a method of wall construction in which branches or thin laths (wattles) were roughly plastered over with mud or clay (daub).

Dentil Small toothlike block, closely spaced in series forming an ornamental band.

Distemper Paint Opaque, water-based paint made with tempera colors, whiting and sizing; also called calcimine paint; a term used mostly in England.

Doric Order One of the five classical orders; developed in Greece. The ornament of the entablature is traditionally thought to have been a metaphor for the beams and supports of timber construction.

Dressing Term used in joinery applied to any molding or finishing.

Dutch Gold, Dutch Metal Metal leaf sometimes used as a less expensive substitute for gold leaf in gilding.

Entablature Horizontal member supported by the column; its three primary divisions are the architrave, frieze and cornice. Each order prescribed specific proportions and ornament for the entablature.

Fascia Plain, flat, horizontal band often used on architraves.

Feather-Edged Tapered edge; an element triangular in section.

Field Upper portion of a wall, between the dado and cornice.

Fielded Panel Raised panel.

Fillet Also known as a listel or annulet; a plain, narrow face used to separate other moldings; one of eight classical molding shapes.

Float Flat tool used by plasterers to smooth or texture plaster surfaces.

Fluting Vertical channels of rounded sections cut in shafts of columns or pilasters.

Fret Also known as a Greek key; incised, painted or applied ornament of thin lines arranged continuously in interlocking rectangular forms.

Frieze Second or middle member of the three primary divisions of the entablature, resting on the architrave and carrying the cornice.

Furring Channel, Furring Strip Wood or metal spacers attached to the structure of a building to provide a level surface for finishes.

Gauging Plaster Specially formulated gypsum plaster, often mixed with lime putty for plaster finish coats.

Gesso Mixture of whiting, glue and gypsum plaster; used to give a plasterlike appearance to other materials and to provide a surface for decorative painting.

Gilding Application of thin sheets of metal, known as metal leaf, to a surface usually prepared with gesso and bole.

Graining Also known as "faux bois"; a simulated wood finish, generally used for inferior woods to create the impression of finer woods.

Guilloche Carved ornament consisting of circular interlaced bands forming a repeating figure.

Half-Lapped Joint at the intersection of two wood members where half of the thickness of each of the two pieces is removed so that they form a flush surface when overlapped.

Ionic Order One of the five classical orders developed in Asia Minor c. 6th century B.C.; column capitals have volutes and the cornice usually has dentils.

Inverted Cymatium Also known as a cyma reversa or reverse ogee; one of eight classical molding shapes.

Joinery Name given to all trim and finishes in architectural woodwork that are framed or fitted together, especially on the interior; distinguished from carpentry, which includes rough framing and timber work. Joinery usually includes stairs, doors, windows and dressings; also wood coverings (e.g., paneling) for rough timber.

Key Plaster anchor formed between and behind openings in lath.

Lath Backing for plaster; made of wood strips, metal sheets, wires or ribs or gypsum board and applied to the structure or furring strips; forms an adhesive or mechanical bond with a plaster base coat.

Marbleizing Also known as "faux marbre"; painted imitation of the color, sheen and veining of marble.

Millwork Includes woodwork such as molding, doors, door frames, window sashes, stairwork and cabinets; now manufactured at a wood-planing mill or woodworking plant.

Modillion Small scrolled bracket or console.

Module Proportional unit of measurement used to give relative dimensions to the parts of the orders; based on the diameter or half a diameter of a column.

Molding Element of construction or decoration used to create varieties of contours and of light and shadow and to indicate and hide joints in masonry, wood and plaster. Molding was generally used to mark the boundary between different features such as the architrave and frieze, or between different members of the same feature, such as the shaft of a column and the capital.

Molding Plaster Type of gypsum plaster specially formulated for casting.

Mortise and Tenon Wood joint where one piece of wood has a rectangular recess called a mortise to receive a rectangular tenon projecting from the end of another piece of wood.

Mud Plaster Used as an interior and exterior surface coating on adobe buildings; composed of clay, sand, water and straw or grass

Muntins Short, vertical intermediate framing members between rails, such as in doors and paneling.

Order Total assembly of parts making up the column and its entablature. The Greek orders were the Doric, Ionic and Corinthian; the Romans added the Tuscan and Composite orders.

Ovolo Also known as an echinus or quarter round; a convex shape, with a quarter-circle profile; one of eight classical molding shapes.

Patera Cup-shaped, antique ornament used in friezes; now refers to any circular ornament applied in relief to a flat surface.

Pediment Triangular gable end of a roof formed by a horizontal cornice line; also used ornamentally. Door and window openings were surmounted by small-scale pediments of many varieties including triangular, broken, segmental and scrolled.

Pedestal Molded block set beneath a column or pilaster divided into three main parts: the base, the dado or die and the surbase, cornice or cap.

Pilaster Engaged pier or column, usually with the details of a column; also used as a decorative element on door casings and mantelpieces.

Plaster Mixture of cementing material (lime, gypsum or portland cement), an aggregate (e.g., sand) and water.

Plinth Block serving as the base of a column or pedestal.

Quirk (or Quirked) Bead Bead with a quirk on one side, as on the edge of a board, so that it appears to be separate from the surface on which it is planed.

Rail Horizontal board used in framing or paneling; usually mortised to receive the tenons of the stiles and grooved for the tongues of the panels.

Reeding Series of convex moldings, the opposite of flutes; sometimes used instead of flutes in columns and pilasters.

Rock Lath Gypsum board lath with surface paper containing gypsum crystals to form a strong bond with plaster.

Rod Also called a straight-edge; used in plastering to straighten plaster surfaces.

Scotia Also known as a trochilos or cove; a rounded concave shape; one of eight classical molding shapes.

Screed Narrow strips of plaster, set in plumbed, level lines; used in plastering to guide the depth of plaster coats.

Shoe Also known as shoe mold or toe mold; the molding at the bottom of the base, whose purpose is to hide the joint between the floor and base if the base or floor shrinks.

Sizing Glue or casein added to water-based paints to increase durability.

Slicker Tool used in plastering to smooth and straighten plaster coats.

Soffit Exposed horizontal underside of any architectural element such as a cornice or lintel.

Stenciling Painted decoration applied through a cut-out template to create a repeating pattern.

Stile Vertical board used in framing or paneling, usually tenoned to fit into the mortises of the rails and grooved for the tongues of panels.

Summer Beam In early 17th- and 18th-century construction, the large wooden beam running from the chimney to the girder in the exterior structural frame.

Torus Convex, semicircular molding; one of eight classical molding shapes.

Thumbnail Bead Quarter-round planed at the edge of a board, recessed slightly from the surface from which it is cut; usually used on the stiles and rails of Georgian doors and fielded wall panels.

Tongue and Groove Method of joing wooden elements; the projecting tongue of one element is inserted into the groove of another element.

Trefoil Three-lobed pattern similar to a cloverleaf; a popular motif in Gothic Revival decoration.

Triglyph Block with vertical channels which is a distinguishing element of the Doric entablature; combined with the mutule above and guttae below, the system represented a masonry metaphor of the features derived from timber construction.

Trim Visible woodwork or molding of a building, including baseboards, moldings and casings, which cover joints, edges and ends of other materials.

Trowel Finish Smooth plaster finish produced by a steel trowel.

Tuscan Order One of the classical orders developed by the Romans, derived from an ancient type of Etruscan temple.

Veneer Thin pieces of wood, usually ornamental, used to cover inferior wood for decorative purposes or to add strength.

Volute Scroll or spiral occurring in Ionic, Corinthian and Composite capitals.

Wainscot Facing, decorative or protective, for the lower portion of a wall; often of boards or panels. The term originally referred to quartered oak, then to boarding or paneling made of it, then sheathing or lining for walls.

Wallboard Rigid sheet composed of wood pulp, gypsum and other materials used for interior wall finishes; early wallboards were of layered fiber, compressed fiber cemented with asphalt, or cardboard.

Whitewash Water-based paint made from ground chalk, salt and lime and sometimes tinted.

Whiting Powdered calcium carbonate pigment, including ground chalk or clay, used in the composition of some paints such as whitewash.

Woodwork Work, often decorative, produced by carpenters and joiners and generally applied to parts of objects or wood structures.

INFORMATION SOURCES

American Association for
   State and Local History
172 2nd Avenue, North
Suite 102
Nashville, Tenn. 37201

American Society of
   Interior Designers
Historic Preservation
   Committee
1430 Broadway
New York, N.Y. 10018

Association for Preservation
   Technology
904 Princess Anne Street
P.O. Box 8178
Fredericksburg, Va. 22404

Campbell Center for Historic
   Preservation Studies
P.O. Box 66
Mount Carroll, Ill. 61053

Decorative Arts Society
c/o Brooklyn Museum
200 Eastern Parkway
Brooklyn, N.Y. 11238

National Council for
   Preservation Education
Heritage Preservation
   Program
Georgia State University
Atlanta, Ga. 30303

National Park Service
Preservation Assistance
   Division
P.O. Box 37127
Washington, D.C. 20013

National Preservation
   Institute
Pension Building
Judiciary Square
Washington, D.C. 20001

**National Trust for
Historic Preservation**
Regional Offices:

Northeast Regional Office
Old City Hall
45 School Street, 4th floor
Boston, Mass. 02108

Mid-Atlantic
   Regional Office
6401 Germantown Avenue
Philadelphia, Pa. 19144

Southern Regional Office
456 King Street
Charleston, S.C. 29403

Midwest Regional Office
53 West Jackson Boulevard
Suite 1135
Chicago, Ill. 60604

Mountains/Plains
   Regional Office
511 16th Street
Suite 700
Denver, Colo. 80202

Texas/New Mexico
   Field Office
500 Main Street
Suite 606
Fort Worth, Tex. 76102

Western Regional Office
One Sutter Street
Suite 707
San Francisco, Calif. 94104

Society for the Preservation
   of New England
   Antiquities
141 Cambridge Street
Boston, Mass. 02114

Society of Architectural
   Historians
1232 Pine Street
Philadelphia, Pa. 19107-5944

The Victorian Society
   in America
219 East 6th Street
Philadelphia, Pa. 19106

FURTHER READING

The following is a list of publications organized by chapter title. Many provided valuable background for the preparation of this book. For anyone interested in further information on any topics discussed here, periodicals such as *Association for Preservation Technology Bulletin, Fine Homebuilding, Fine Woodworking, Historic Preservation, The Old-House Journal* and *Traditional Building* give insightful and thorough discussion of many issues affecting the repair and rehabilitation of walls and molding.

American Interior Styles

Architects' Emergency Committee. *Great Georgian Houses of America.* Books I and II. New York: Kalkhoff Press, 1933–37.

Benjamin, Asher. *The American Builder's Companion.* 6th ed. 1827. Reprint. New York: Dover, 1969.

Bicknell, A. J. *Cottage and Villa Architecture.* New York: 1878.

————. *Detail, Cottage and Constructive Architecture.* New York: 1873.

Bowyer, Jack. *Handbook of Building Crafts in Conservation.* London: Hutchinson & Co., 1981.

Bunting, Bainbridge. *Early Architecture of New Mexico.* Albuquerque: University of New Mexico Press, 1976.

————. *Taos Adobes.* Santa Fe: Museum of New Mexico Press, 1964.

Clute, Eugene. *The Treatment of Interiors.* New York: Pencil Points Press, 1926.

Coffin, Lewis A., Jr., and Arthur C. Holden. *Brick Architecture of the Colonial Period in Maryland and Virginia.* 1919. Reprint. New York: Dover, 1970.

Cummings, Abbott Lowell. *The Framed Houses of the Massachusetts Bay, 1625–1725.* Cambridge, Mass.: Belknap Press, 1979.

Davidson, E. A. *A Practical Manual of Housepainting, Graining, Marbling, and Sign Writing, Containing Full Information on the Processes of House Painting in Oil and Distemper...* London: 1900.

Downing, Andrew Jackson. *The Architecture of Country Houses.* 1850. Reprint. New York: Dover, 1969.

Eastlake, Charles. *Hints on Household Taste in Furniture, Upholstery, and Other Details.* 1868. Reprint. New York: Dover, 1969.

Eight Periods and Their Modern Adaptation Together with a Chapter on the Spanish Style. Newark, N.J.: Murphy Varnish Company, 1925.

Elliott, Charles Wyllys. *The Book of American Interiors.* Boston: James R. Osgood, 1876

Englund, John H. "An Outline of the Development of Wood Moulding Machinery," *APT Bulletin* 10, no. 4, pp. 20–46.

Forman, H. Chandlee. *Early Manor and Plantation Houses of Maryland.* 2nd ed. Baltimore: Bodine & Associates, with Maclay & Associates, 1982.

————. *Maryland Architecture.* Cambridge, Md.: Tidewater Publishers, 1968.

————. *Tidewater Maryland Architecture and Gardens.* New York: Bonanza Books, 1956.

Gilmore, Andrea. "Dating Architectural Moulding Profiles — A Study of Eighteenth and Nineteenth Century Moulding Plane Profiles in New England," *APT Bulletin* 10, no. 2, pp. 91–117.

Gottfried, Herbert, and Jan Jennings. *American Vernacular Design 1870–1940.* New York: Van Nostrand Reinhold, 1985.

Gowans, Alan. *The Comfortable House: North American Suburban Architecture 1890–1930.* Cambridge, Mass.: MIT Press, 1986.

————. *Images of American Living.* New York: Harper & Row, Icon Editions, 1976.

Grow, Lawrence, ed. *The Old House Book Series: Living Rooms and Parlors; Bedrooms; Kitchens and Dining Rooms; Outdoor Living Spaces.* New York: Warner Books, 1980–.

Hamlin, Talbot. *Greek Revival Architecture in America.* New York: Dover, 1963.

Hannaford, Donald R., and Revel Edwards. *Spanish Colonial or Adobe Architecture of California 1801–1850.* New York: Architectural Book Publishing Company, 1931.

Harris, Cyril M., ed. *Dictionary of Architecture and Construction.* New York: McGraw-Hill, 1975.

Hodgson, Fred T. *Mortars, Plasters, Stuccos.* 1906. Rev. Chicago: Frederick J. Drake, 1916.

Holly, Henry Hudson. *Modern Dwellings in Town and Country Adapted to American Wants and Climate.* New York: 1878.

————. *Victorian Country Seats and Modern Dwellings.* 1865. Reprint. Watkins Glen, N.Y.: American Life Foundation, 1977.

Isham, Norman Morrison. *Early American Houses.* 1928. Reprint. Watkins Glen, N.Y.: American Life Foundation, 1968.

————. *A Glossary of Colonial Architectural Terms.* 1939. Reprint. Watkins Glen, N.Y.: American Life Foundation, 1968.

Jennings, Jan, and Herbert Gottfried. *American Vernacular Interior Architecture (1870–1940).* New York: Van Nostrand Reinhold, 1988.

Kelly, J. Frederick. *Early Domestic Architecture of Connecticut.* 1924. Reprint. New York: Dover, 1963.

Kettell, Russell Hawes. *Early American Rooms: 1650–1858.* 1936. Reprint. New York: Dover, 1967.

Lafever, Minard. *The Beauties of Modern Architecture.* New York: 1835.

————. *The Modern Builder's Guide.* 1833. Reprint. New York: Dover, 1969.

Lancaster, Clay. *The American Bungalow 1880–1930.* New York: Abbeville Press, 1985.

————. *Photographs of New York Interiors at the Turn of the Century.* Photographs by Joseph Byron. New York: Dover, 1976.

Late Victorian Architectural Details: Combined Book of Sash, Doors, Blinds, Mouldings, Etc. Reprint. Watkins Glen, N.Y.: American Life Foundation, 1978.

Little, Nina Fletcher. *American Decorative Wall Painting, 1700–1850.* New York: Dutton, 1972.

Loth, Calder, and Julius Toursdale Sadler, Jr. *The Only Proper Style: Gothic Architecture in America.* Boston: New York Graphic Society, 1975.

Loudon, John Claudius. *An Encyclopedia of Cottage, Farm and Villa Architecture and Furniture.* New York: 1869.

Mayhew, Edgar deN., and Minor Myers, Jr. *A Documentary History of American Interiors from the Colonial Era to 1915.* New York: Charles Scribner's Sons, 1980.

Mullins, Lisa C., ed. *Architectural Treasures of Early Amer-*

ica. Series. Harrisburg, Pa.: National Historical Society, 1987–.

Newcomb, Rexford. *The Colonial and Federal House.* Philadelphia and London: J. B. Lippincott, 1933.

Palladio, Andrea. *The Four Books of Architecture.* 1738. Reprint. New York: Dover, 1965.

Pierce, Donald C., and Hope Alswang. *American Interiors: New England and the South. Period Rooms at the Brooklyn Museum.* New York: Universe, 1983.

Ranlett, William H. *The Architect: A Series of Original Designs for Domestic and Ornamental Cottages.* New York: 1849.

Sexton, R. W. *Spanish Influence on American Architecture and Decoration.* New York: Brentano's, 1927.

Smith, Mary Ann. *Gustav Stickley, The Craftsman.* Syracuse, N.Y.: Syracuse University Press, 1983.

Stickley, Gustav. *Craftsman Homes.* 1909. Reprint. New York: Dover, 1979.

————. *More Craftsman Homes.* 1912. Reprint. New York: Dover, 1982.

Summerson, John. *The Classical Language of Architecture.* Cambridge, Mass.: MIT Press, 1963.

Vandal, Norman. "Period Moldings," *Fine Homebuilding*, April-May 1984, pp. 59–63.

Vaux, Calvert. *Villas and Cottages.* 1857. Reprint. New York: Dover, 1970.

Victorian Design Book: A Complete Guide to Victorian House Trim. 1904. Reprint. Ottawa, Ontario: Lee Valley Tools, 1984.

Ware, William. *The American Vignola.* New York: W. W. Norton, 1977.

Weil, Martin Eli. "Interior Details in Eighteenth Century Architectural Books," *APT Bulletin* 10, no. 4, pp. 47–66.

Wharton, Edith, and Ogden Codman, Jr. *The Decoration of Houses.* 1897. Reprint. New York: W. W. Norton, 1978.

Wheeler, Gervase. *Homes for People.* New York: 1855.

Whiton, Sherrill. *Elements of Interior Decoration.* Chicago, Philadelphia and New York: J. B. Lippincott, 1957.

Winkler, Gail Caskey, and Roger W. Moss. *Victorian Interior Decoration: American Interiors 1830–1900.* New York: Henry Holt, 1986.

Winter, Robert. *The California Bungalow.* Los Angeles: Hennessey & Ingalls, 1980.

Woodward, George, and Edward G. Thompson. *Woodward's National Architect; Containing 100 Original Designs, Plans and Details, to Working Scale, for the Practical Construction of Dwelling Houses for the Country, Suburb and Village.* 1869. Reprint. New York: DaCapo Press, 1975.

Zolotow, Maurice. "Victorian Dream," *American Preservation*, vol. 1, no. 1, pp. 34–41.

Preserving the Character of Your Interior

Planning for Rehabilitation

Bullock, Orin M., Jr. *The Restoration Manual: An Illustrated Guide to the Preservation and Restoration of Old Buildings.* Norwalk, Conn.: Silvermine Publishing, 1966.

"Danger: Restoration May Be Hazardous to Your Health," *The Old-House Journal*, vol. IV, no. 5, pp. 9–11.

"Guidelines for Rehabilitating Old Buildings," *The Old-House Journal*, vol. V, no.1, pp. 8–11.

Hutchins, Nigel. *Restoring Old Houses.* Toronto: Van Nostrand Reinhold, 1980.

Litchfield, Michael W. *Renovation: A Complete Guide.* New York: John Wiley & Sons, 1982

London, Mark. *Masonry: How to Care for Old and Historic Brick and Stone.* Washington, D.C.: Preservation Press, 1988.

Maddex, Diane, ed. *All About Old Buildings: The Whole Preservation Catalog*. Washington, D.C.: Preservation Press, 1985.

Morton, W. Brown, III, Gary L. Hume and Kay D. Weeks. *The Secretary of the Interior's Standards for Rehabilitation and Guidelines for Rehabilitating Historic Buildings*. Rev. ed. Washington, D.C.: Technical Preservation Services, U.S. Department of the Interior, 1990.

Powers, Alice. "Almost the Real Thing," *Historic Preservation*, May-June 1987, pp. 22–25.

Prentice, Helaine Kaplan, and Blair Prentice. *Rehab Right*. Berkeley, Calif.: Ten Speed Press, 1986.

Seale, William. *Recreating the Historic House Interior*. Nashville: American Association for State and Local History, 1979.

Stephen, George. *New Life for Old Houses*. 1972. Reprint. Washington, D.C.: Preservation Press, 1989.

Stieglitz, Maria. "Safety First," *Historic Preservation*, July-August 1989, pp. 10–13.

Stoddard, Brooke C. "How the Experts Solve Problems," *Historic Preservation*, January-February 1987, pp. 46–51.

Weaver, Martin E. "The Investigation and Recording of Moulding Profiles," *APT Bulletin* 10, no. 4, pp. 88–92.

Rehabilitating Woodwork

"After You Strip — Before You Finish," *The Old-House Journal*, vol. XV, no. 1, pp. 43–47.

Alvarez, Mark. "Stripping Trim," *Fine Homebuilding*, August-September 1983, pp. 66–69.

Berney, Bruce R. "Removing Woodwork for Paint Stripping," *The Old-House Journal*, February 1978, p. 13.

Bock, Gordon H. "Removing Interior Woodwork," *The Old-House Journal*, June 1985, pp. 108–11.

Brown, Burton. "Making Wood Mouldings the Old Way," *The Old-House Journal*, April 1976, pp. 4–11.

"Commercial Paint Stripping," *The Old-House Journal*, July-August 1988, pp. 29–33.

Finishing Manual. Ottawa, Ontario: Lee Valley Tools, 1987.

Frank, George. "Stains, Dyes and Pigments," *Fine Woodworking*, September 1978, pp. 58–59.

Gibbia, S. W. *Wood Finishing and Refinishing*. Englewood Cliffs, N.J.: Prentice Hall, 1986.

Johnson, Ed. *Old House Woodwork Restoration*. Englewood Cliffs, N.J.: Prentice-Hall, 1983.

Labine, Clem. "Dealing with Calcimine Paint," *The Old-House Journal*, May 1976, pp. 2–3.

————. "Dip-Stripping to Remove Paint," *The Old-House Journal*, August 1982, pp. 157–60.

————. "Restoring Clear Finishes," *The Old-House Journal*, November 1982, p. 221.

————. "Remedies for 'Dark, Ugly' Woodwork," *The Old-House Journal*, September 1976, p. 5.

Mussey, Robert D. "Early Varnishes," *Fine Woodworking*, July-August 1982, pp. 54–57.

————. *The First American Furniture Finisher's Manual*. 1827. Reprint. New York: Dover, 1987.

————. "Old Finishes," *Fine Homebuilding*, March-April 1982, pp. 71–79.

Newell, Arthur D., "Finishing Materials," *Fine Woodworking*, July-August 1979, pp. 72–75.

O'Donnell, Bill. "Flow-On Paint Stripping," *The Old-House Journal*, July-August 1986, pp. 279–81.

Oughton, Frederick. *The Complete Manual of Wood Finishing*. New York: Stein and Day, 1985.

Poore, Patricia. "The Basics of Stripping Paint," *The Old-House Journal*, January-February 1988, pp. 38–43.

"Refinishing Clinic," *The Old-House Journal*, vol. IX, no. 10, pp. 229–31.

Williams, Donald C. "Shellac Finishing," *Fine Woodworking*, July-August 1988, pp. 56–59.

Preserving Plasterwork

"The Art of Getting Plastered — Duplicating Plaster Cornices," *The Old-House Journal*, vol. II, no. 2, pp. 7–9.

Flaharty, David. "Ornamental Plaster Restoration," *Fine Homebuilding*, January 1990, pp. 38–42.

Garrison, John Mark. "Casting Decorative Plaster," *The Old-House Journal*, vol. XIII, no. 9, pp. 186–89.

—————. "Decorative Plaster: Running Cornices," *The Old-House Journal*, vol. XII, no. 10, pp. 214–19.

—————. "Running Plaster Mouldings," *The Old-House Journal*, vol. XII, no. 7, pp. 136–141.

Leeke, John. "Saving Irreplaceable Plaster," *The Old-House Journal*, vol. XV, no. 6, pp. 51–55.

MacDonald, Marylee. *Plaster.* Old House Restoration No.1. Champaign: University of Illinois, 1984.

—————. *Repairing Historic Flat Plaster — Walls and Ceilings.* Preservation Brief no. 21. Washington, D.C.: Technical Preservation Services, U.S. Department of the Interior, 1989.

McKee, Harley J. *Introduction to Early American Masonry: Stone, Brick, Mortar and Plaster.* Washington, D.C.: Preservation Press, 1973.

Nelson, Lee H. *Preservation of Historic Adobe Buildings,* Preservation Brief no. 5. Washington, D.C.: Technical Preservation Services, U.S. Department of the Interior, 1978.

Phillips, Morgan W. "Adhesives for the Reattachment of Loose Plaster," *APT Bulletin* 12, no.2, pp. 37–63.

Poore, Patricia. "The Basics of Plaster Repair," *The Old-House Journal*, March-April 1988, pp. 29–33.

Snyder, Tim. "Veneer Plaster," *Fine Homebuilding*, June-July 1983, pp. 72–74.

Todaro, John. "Molding and Casting Materials," *Fine Homebuilding*, February-March 1981, pp. 36–39.

Van Den Branden, F., and Thomas L. Hartsell. *Plastering Skills.* Homewood, Ill.: American Technical Publishers, 1984.

Reviving Decorative Painted Finishes

Bishop, Adele, and Cile Lord. *The Art of Decorative Stencilling.* New York: Penguin, 1978.

Exterior Decoration. The Atheneum Library of Nineteenth Century America. Philadelphia: The Atheneum of Philadelphia, 1976.

Fraser, Bridget. *Stencilling: A Design and Source Book.* New York: Henry Holt, 1987.

Innes, Jocasta. *Paint Magic: The Complete Guide to Decorative Finishes.* New York: Van Nostrand Reinhold, 1981.

Jansen, James, "How to Glaze Walls & Ceilings," *The Old-House Journal*, January-February 1988, pp. 27–33.

Mosca, Matthew. "Historic Paint Research: Determining the Original Colors," *The Old-House Journal*, vol. IX, no. 4, pp. 81–83.

O'Neil, Isabel. *The Art of the Painted Finish for Furniture and Decoration.* New York: William Morrow, 1971.

Penn, T. Z. "Decorative and Painted Finishes," *APT Bulletin* 16, no. 1, pp. 3–46.

Waring, Janet. *Early American Wall Stencils.* New York: William R. Scott, 1937.

Weinstein, Nat. "The Art of Graining," *The Old-House Journal*, vol. VI, no. 12, p. 133.

Welsh, Frank S. "A Methodology for Exposing and Preserving Architectural Graining," *APT Bulletin* 8, no. 2, p. 71.

—————. "Paint Analysis," *APT Bulletin* 14, no. 4, pp. 29–30.

—————. "Paint and Color Restoration," *The Old-House Journal*, vol. III, no. 8, p. 1.

INDEX

Page numbers in italics refer to illustrations and captions.

AUTHOR

Natalie Shivers, an architectural historian and designer with the architectural firm Levin and Associates in Los Angeles, is currently in charge of the rehabilitation of the historic Bradbury Building, Grand Central Market and Million Dollar Theater Building. From 1985 to 1989 she worked with Hardy Holzman Pfeiffer Associates on the rehabilitation and expansion of the Los Angeles Central Library and the Los Angeles City Hall. Her preservation experience includes stints with Savannah Landmarks, the Irish Georgian Society and the Maryland Historical Trust, where she served as architectural historian and preservation coordinator. Shivers has also written *Those Old Placid Rows: The Development and Aesthetic of the Baltimore Rowhouse* (Maclay & Associates).

Ornamental plaster ceiling, 1916, at 1785 Massachusetts Avenue, N.W., Washington, D.C., headquarters of the National Trust for Historic Preservation. (Robert Grove)